OFFICIAL SQA SPECIMEN QUESTION PAPER AND HODDER GIBSON MODEL QUESTION PAPERS WITH ANSWERS

NATIONAL 5

PHYSICS

2013 Specimen Question Paper & 2013 Model Papers

HODDER GIBSON
LEARN MORE

This book contains the official 2013 SQA Specimen Question Paper for National 5 Physics, with associated SQA approved answers modified from the official marking instructions that accompany the paper.

In addition the book contains model practice papers, together with answers, plus study skills advice. These papers, some of which may include a limited number of previously published SQA questions, have been specially commissioned by Hodder Gibson, and have been written by experienced senior teachers and examiners in line with the new National 5 syllabus and assessment outlines, Spring 2013. This is not SQA material but has been devised to provide further practice for National 5 examinations in 2014 and beyond.

Hodder Gibson is grateful to the copyright holders, as credited on the final page of the Answer Section, for permission to use their material. Every effort has been made to trace the copyright holders and to obtain their permission for the use of copyright material. Hodder Gibson will be happy to receive information allowing us to rectify any error or omission in future editions.

Hachette UK's policy is to use papers that are natural, renewable and recyclable products and made from wood grown in sustainable forests. The logging and manufacturing processes are expected to conform to the environmental regulations of the country of origin.

Orders: please contact Bookpoint Ltd, 130 Park Drive, Abingdon, Oxon OX14 4SE. Telephone: (44) 01235 827720. Fax: (44) 01235 400454. Lines are open 9.00–5.00, Monday to Saturday, with a 24-hour message answering service. Visit our website at www.hoddereducation.co.uk. Hodder Gibson can be contacted direct on: Tel: 0141 848 1609; Fax: 0141 889 6315; email: hoddergibson@hodder.co.uk

This collection first published in 2013 by
Hodder Gibson, an imprint of Hodder Education,
An Hachette UK Company
2a Christie Street
Paisley PA1 1NB

BrightRED Hodder Gibson is grateful to Bright Red Publishing Ltd for collaborative work in preparation of this book and all SQA Past Paper and National 5 Model Paper titles 2013.

Typeset by PDQ Digital Media Solutions Ltd, Bungay, Suffolk NR35 1BY

Printed in the UK

A catalogue record for this title is available from the British Library

ISBN: 978-1-4718-0227-0

3 2

2014

Introduction

Study Skills – what you need to know to pass exams!

Pause for thought

Many students might skip quickly through a page like this. After all, we all know how to revise. Do you really though?

Think about this:

"IF YOU ALWAYS DO WHAT YOU ALWAYS DO, YOU WILL ALWAYS GET WHAT YOU HAVE ALWAYS GOT."

Do you like the grades you get? Do you want to do better? If you get full marks in your assessment, then that's great! Change nothing! This section is just to help you get that little bit better than you already are.

There are two main parts to the advice on offer here. The first part highlights fairly obvious things but which are also very important. The second part makes suggestions about revision that you might not have thought about but which WILL help you.

Part 1

DOH! It's so obvious but …

Start revising in good time

Don't leave it until the last minute – this will make you panic.

Make a revision timetable that sets out work time AND play time.

Sleep and eat!

Obvious really, and very helpful. Avoid arguments or stressful things too – even games that wind you up. You need to be fit, awake and focused!

Know your place!

Make sure you know exactly **WHEN and WHERE** your exams are.

Know your enemy!

Make sure you know what to expect in the exam.

How is the paper structured?

How much time is there for each question?

What types of question are involved?

Which topics seem to come up time and time again?

Which topics are your strongest and which are your weakest?

Are all topics compulsory or are there choices?

Learn by DOING!

There is no substitute for past papers and practice papers – they are simply essential! Tackling this collection of papers and answers is exactly the right thing to be doing as your exams approach.

Part 2

People learn in different ways. Some like low light, some bright. Some like early morning, some like evening / night. Some prefer warm, some prefer cold. But everyone uses their BRAIN and the brain works when it is active. Passive learning – sitting gazing at notes – is the most INEFFICIENT way to learn anything. Below you will find tips and ideas for making your revision more effective and maybe even more enjoyable. What follows gets your brain active, and active learning works!

Activity 1 – Stop and review

Step 1

When you have done no more than 5 minutes of revision reading STOP!

Step 2

Write a heading in your own words which sums up the topic you have been revising.

Step 3

Write a summary of what you have revised in no more than two sentences. Don't fool yourself by saying, 'I know it but I cannot put it into words'. That just means you don't know it well enough. If you cannot write your summary, revise that section again, knowing that you must write a summary at the end of it. Many of you will have notebooks full of blue/black ink writing. Many of the pages will not be especially attractive or memorable so try to liven them up a bit with colour as you are reviewing and rewriting. **This is a great memory aid, and memory is the most important thing.**

Activity 2 — Use technology!

Why should everything be written down? Have you thought about 'mental' maps, diagrams, cartoons and colour to help you learn? And rather than write down notes, why not record your revision material?

What about having a text message revision session with friends? Keep in touch with them to find out how and what they are revising and share ideas and questions.

Why not make a video diary where you tell the camera what you are doing, what you think you have learned and what you still have to do? No one has to see or hear it but the process of having to organise your thoughts in a formal way to explain something is a very important learning practice.

Be sure to make use of electronic files. You could begin to summarise your class notes. Your typing might be slow but it will get faster and the typed notes will be easier to read than the scribbles in your class notes. Try to add different fonts and colours to make your work stand out. You can easily Google relevant pictures, cartoons and diagrams which you can copy and paste to make your work more attractive and **MEMORABLE**.

Activity 3 – This is it. Do this and you will know lots!

Step 1

In this task you must be very honest with yourself! Find the SQA syllabus for your subject (www.sqa.org.uk). Look at how it is broken down into main topics called MANDATORY knowledge. That means stuff you MUST know.

Step 2

BEFORE you do ANY revision on this topic, write a list of everything that you already know about the subject. It might be quite a long list but you only need to write it once. It shows you all the information that is already in your long-term memory so you know what parts you do not need to revise!

Step 3

Pick a chapter or section from your book or revision notes. Choose a fairly large section or a whole chapter to get the most out of this activity.

With a buddy, use Skype, Facetime, Twitter or any other communication you have, to play the game "If this is the answer, what is the question?". For example, if you are revising Geography and the answer you provide is "meander", your buddy would have to make up a question like "What is the word that describes a feature of a river where it flows slowly and bends often from side to side?".

Make up 10 "answers" based on the content of the chapter or section you are using. Give this to your buddy to solve while you solve theirs.

Step 4

Construct a wordsearch of at least 10 X 10 squares. You can make it as big as you like but keep it realistic. Work together with a group of friends. Many apps allow you to make wordsearch puzzles online. The words and phrases can go in any direction and phrases can be split. Your puzzle must only contain facts linked to the topic you are revising. Your task is to find 10 bits of information to hide in your puzzle but you must not repeat information that you used in Step 3. DO NOT show where the words are. Fill up empty squares with random letters. Remember to keep a note of where your answers are hidden but do not show your friends. When you have a complete puzzle, exchange it with a friend to solve each other's puzzle.

Step 5

Now make up 10 questions (not "answers" this time) based on the same chapter used in the previous two tasks. Again, you must find NEW information that you have not yet used. Now it's getting hard to find that new information! Again, give your questions to a friend to answer.

Step 6

As you have been doing the puzzles, your brain has been actively searching for new information. Now write a NEW LIST that contains only the new information you have discovered when doing the puzzles. Your new list is the one to look at repeatedly for short bursts over the next few days. Try to remember more and more of it without looking at it. After a few days, you should be able to add words from your second list to your first list as you increase the information in your long-term memory.

FINALLY! Be inspired...

Make a list of different revision ideas and beside each one write **THINGS I HAVE** tried, **THINGS I WILL** try and **THINGS I MIGHT** try. Don't be scared of trying something new.

And remember – "FAIL TO PREPARE AND PREPARE TO FAIL!"

National 5 Physics

The exam

Duration: 2 hours
Total marks: 110

20 marks are awarded for 20 **multiple-choice questions** – completed on an answer grid.

90 marks are awarded for **written answers** – completed in the space provided after each question or on graph paper.

Approximately one third of the 110 marks are allocated to questions from each unit.

The National 5 Physics course consists of **three units**:

- Unit 1 – Electricity and Energy
- Unit 2 – Waves and Radiation
- Unit 3 – Dynamics and Space

General exam advice

There are 110 marks in total, and you have two hours to complete the paper. This works out at just over one minute per mark, so a 10 mark question would take roughly 11 minutes.

Be aware of how much time you spend on each question. For example, DO NOT spend 10 minutes on a question worth only three marks, especially when you haven't completed the rest of the questions – you can always return to the question later if there's time.

The best method for getting used to National 5 exam questions is to attempt as many exam type questions as possible, **and check your answers**. If you find a wrong answer, **find out why it is wrong** and then try similar questions until you can answer them correctly.

Specific exam advice

Advice for answering multiple-choice questions (Section 1) (20 marks)

Each question has five possible choices of answers. **Only one answer is correct.**

Multiple-choice questions are designed to test a range of skills, e.g.

- knowledge and understanding of the course
- using equations
- selecting correct statements from a list
- selecting and analysing information from a diagram.

It is important to **practise** as many **multiple-choice questions** as possible, to get used to the 'style' and types of questions.

Do not try to work out all of the answers to multiple-choice questions in your head. Instead, when the question is complicated, write down notes and work on scrap paper (provided by the invigilator) or use the blank pages at the end of the question paper.

Do not use the answer grid for working, and remember to cross out your multiple-choice rough working when you have finished.

You can also make notes beside the actual question if this helps, but **not** on the answer grid.

Advice for answering written questions (Section 2) (90 marks)

These questions test several different skills.

The majority of these marks test your **knowledge and understanding** of the course.

There are also questions which test different skills, like selecting information, analysing information, predicting results, and commenting on experimental results.

There are usually around 12–14 questions in Section 2. There are different types of questions, which include:

- Questions testing your **knowledge of the course**, sometimes applied to particular applications. More than half of the 90 marks in Section 2 are for this type of question.

- Questions (usually a maximum of two) involving **physics content not in the course**, but explained in the question, usually including an equation which you are asked to use with data.

- A question testing your **scientific reading skills**, where you will be asked about a scientific report or passage. The question might include a calculation.

- **'Open-ended' questions** (a maximum of 2 per exam, three marks each), which usually discuss a physics phenomenon and ask you to explain it using your knowledge of physics. You have to think about the issue and try to give a step by step answer – there may be more than one area of physics used to answer this type of question. These questions allow you to use your knowledge and problem solving skills. Be careful not to spend longer than necessary on these three mark questons.

- Questions testing practical skills usually based on **practical or experimental work**, which may have tables of results or graphs (or both) which have to be used to obtain information needed to answer the question. You could be asked to identify a problem with the results, or to suggest an improvement to the experiment.

Things to remember when answering questions

Using equations

More than half of the total marks awarded in Section 2 are for being able to calculate answers using an equation (relationship) from the **'Relationship Sheet'** which is supplied with the exam paper.

These questions are usually worth three marks. To obtain the full three marks for these questions, your final answer must be correct.

There are three separate marks awarded for the stages of the working:

- Write down the correct equation needed to calculate the answer from the Relationship Sheet – **1 mark**.

- Show that the correct values are substituted into the equation – **1 mark**.

- Show the final answer, including the correct unit – **1 mark**.

If the unit is wrong or missing, you will lose the final mark!

Other important areas to remember and practise are:

Units

The units of measurement in the National 5 physics course are based on the International System of units. Make sure that you use the correct unit following a calculation in your final answer.

Prefixes

A prefix produces a multiple of the unit in powers of ten, e.g. 10^{-6} is 0·000001. It is named 'micro' and has the symbol 'μ'. Make sure to practise and get used to all prefixes.

Scientific notation

This is used in the exam to write very large or very small numbers, to avoid writing or using strings of numbers in an answer or calculation.

You need to be familiar with how to enter and use numbers in scientific notation on **your** calculator – make sure that you have used your calculator often before the exam to get used to it.

Significant figures

When calculating a value using an equation, take care not to give too many significant figures in the final answer. If there are intermediate steps in a calculation, you can keep numbers in your calculator which have too many significant figures. You should always round your answer to give no more than the smallest number of significant figures which appear in the data given in the question.

$$\text{E.g. } \frac{42 \cdot 74}{2 \cdot 59} = 16 \cdot 5019305$$

If the smallest number of significant figures relating to the data used from the question was three, then round this answer to 16·5.

Examples:

- 20 has 1 significant figure
- 40·0 has 3 significant figures
- 0·000604 has 3 significant figures
- $4 \cdot 30 \times 10^4$ has 3 significant figures
- 6200 has 2 significant figures

Good luck!

Remember that the rewards for passing National 5 Physics are well worth it! Your pass will help you get the future you want for yourself. In the exam, be confident in your own ability. If you are not sure how to answer a question, trust your instincts and just give it a go anyway. Keep calm and don't panic! GOOD LUCK!

NATIONAL 5

2013 Specimen
Question Paper

National Qualifications
SPECIMEN ONLY

SQ35/N5/01

Physics
Section 1—Questions

Date — Not applicable

Duration — 2 hours

Instructions for completion of Section 1 are given on Page two of the question paper SQ35/N5/02.

Record your answers on the grid on Page three of your answer booklet

Do NOT write in this booklet.

Before leaving the examination room you must give your answer booklet to the Invigilator. If you do not, you may lose ALL the marks for this paper.

DATA SHEET

Speed of light in materials

Material	Speed in $m\,s^{-1}$
Air	$3\cdot0 \times 10^8$
Carbon dioxide	$3\cdot0 \times 10^8$
Diamond	$1\cdot2 \times 10^8$
Glass	$2\cdot0 \times 10^8$
Glycerol	$2\cdot1 \times 10^8$
Water	$2\cdot3 \times 10^8$

Speed of sound in materials

Material	Speed in $m\,s^{-1}$
Aluminium	5200
Air	340
Bone	4100
Carbon dioxide	270
Glycerol	1900
Muscle	1600
Steel	5200
Tissue	1500
Water	1500

Gravitational field strengths

	Gravitational field strength on the surface in $N\,kg^{-1}$
Earth	9·8
Jupiter	23
Mars	3·7
Mercury	3·7
Moon	1·6
Neptune	11
Saturn	9·0
Sun	270
Uranus	8·7
Venus	8·9

Specific heat capacity of materials

Material	Specific heat capacity in $J\,kg^{-1}\,{}^{\circ}C^{-1}$
Alcohol	2350
Aluminium	902
Copper	386
Glass	500
Ice	2100
Iron	480
Lead	128
Oil	2130
Water	4180

Specific latent heat of fusion of materials

Material	Specific latent heat of fusion in $J\,kg^{-1}$
Alcohol	$0\cdot99 \times 10^5$
Aluminium	$3\cdot95 \times 10^5$
Carbon Dioxide	$1\cdot80 \times 10^5$
Copper	$2\cdot05 \times 10^5$
Iron	$2\cdot67 \times 10^5$
Lead	$0\cdot25 \times 10^5$
Water	$3\cdot34 \times 10^5$

Melting and boiling points of materials

Material	Melting point in °C	Boiling point in °C
Alcohol	−98	65
Aluminium	660	2470
Copper	1077	2567
Glycerol	18	290
Lead	328	1737
Iron	1537	2737

Specific latent heat of vaporisation of materials

Material	Specific latent heat of vaporisation in $J\,kg^{-1}$
Alcohol	$11\cdot2 \times 10^5$
Carbon Dioxide	$3\cdot77 \times 10^5$
Glycerol	$8\cdot30 \times 10^5$
Turpentine	$2\cdot90 \times 10^5$
Water	$22\cdot6 \times 10^5$

Radiation weighting factors

Type of radiation	Radiation weighting factor
alpha	20
beta	1
fast neutrons	10
gamma	1
slow neutrons	3

SECTION 1

1. 1 volt is equivalent to

   A   1 ampere per watt

   B   1 coulomb per second

   C   1 joule per coulomb

   D   1 joule per second

   E   1 watt per second.

2. A conductor carries a current of $4 \cdot 0$ mA for 250 s.

   The total charge passing a point in the conductor is

   A $1 \cdot 6 \times 10^{-5}$ C

   B $1 \cdot 0$ C

   C $62 \cdot 5$ C

   D $1 \cdot 0 \times 10^{3}$ C

   E $6 \cdot 25 \times 10^{4}$ C.

3. A ball is released from rest and allowed to roll down a curved track as shown.

   The mass of the ball is $0 \cdot 50$ kg.

   The maximum height reached on the opposite side of the track is $0 \cdot 20$ m lower than the height of the starting point.

   The amount of energy lost is

   A $0 \cdot 080$ J

   B $0 \cdot 10$ J

   C $0 \cdot 98$ J

   D $2 \cdot 9$ J

   E $3 \cdot 9$ J.

4. In the circuit shown, the current in each resistor is different.

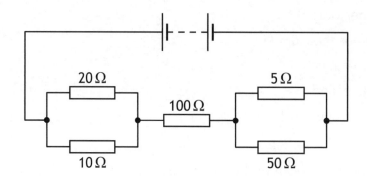

In which resistor is the current smallest?

A   5 Ω

B   10 Ω

C   20 Ω

D   50 Ω

E   100 Ω

5. Three resistors are connected as shown.

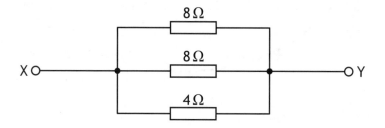

The resistance between X and Y is

A   0·08 Ω

B   0·5 Ω

C   2 Ω

D   13 Ω

E   20 Ω.

6. A bicycle pump is sealed at one end and the piston pushed until the pressure of the trapped air increases to $4 \cdot 00 \times 10^5$ Pa.

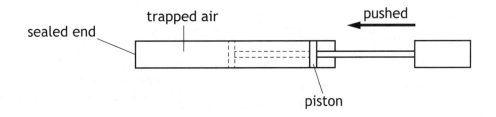

The area of the piston compressing the air is $5 \cdot 00 \times 10^{-4}$ m^2.

The force that the trapped air exerts on the piston is

A $1 \cdot 25 \times 10^{-9}$ N

B $8 \cdot 00 \times 10^{-1}$ N

C $2 \cdot 00 \times 10^{2}$ N

D $8 \cdot 00 \times 10^{8}$ N

E $2 \cdot 00 \times 10^{10}$ N.

7. Which of the following diagrams shows the best method for an experiment to investigate the relationship between pressure and temperature for a fixed mass of gas?

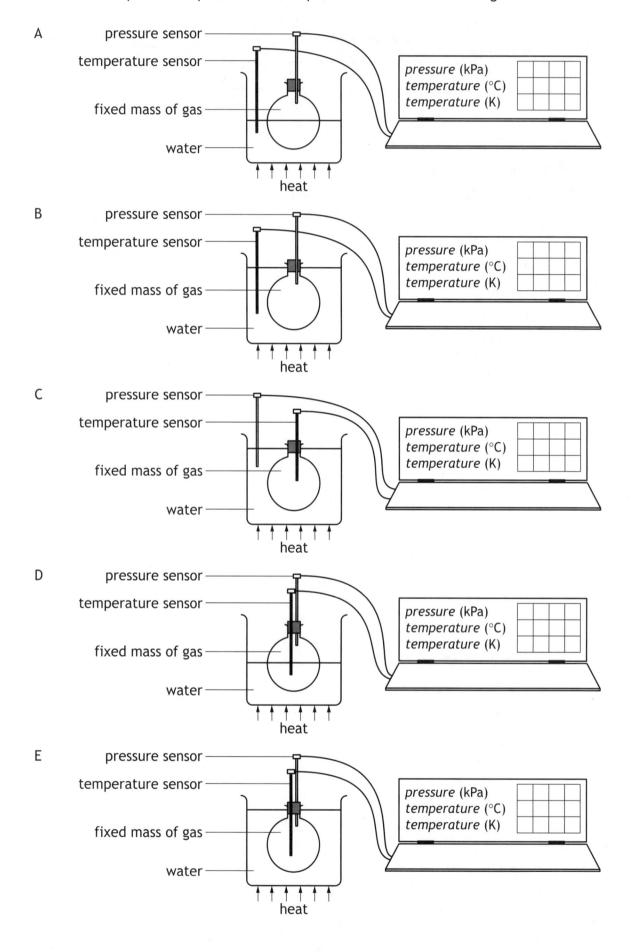

8. A fixed mass of gas is trapped inside a sealed container. The volume of the gas is slowly changed. The temperature of the gas remains constant.

Which graph shows how the pressure p of the gas varies with the volume V?

A
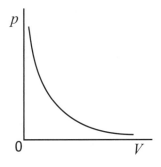

B
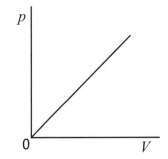

C
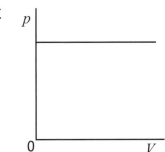

D
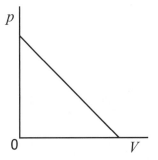

E
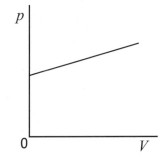

9. A student writes the following statements about electromagnetic waves.

    I   Electromagnetic waves all travel at the same speed in air.

    II Electromagnetic waves all have the same frequency.

    III Electromagnetic waves all transfer energy.

    Which of these statements is/are correct?

    A   I only

    B   II only

    C   I and III only

    D   II and III only

    E   I, II and III

10. A satellite orbiting the Earth transmits television signals to a receiver.
    The signals take a time of 150 ms to reach the receiver.
    The distance between the satellite and the receiver is

    A $2 \cdot 0 \times 10^6$ m

    B $2 \cdot 25 \times 10^7$ m

    C $4 \cdot 5 \times 10^7$ m

    D $2 \cdot 0 \times 10^9$ m

    E $4 \cdot 5 \times 10^{10}$ m.

11. A wave machine in a swimming pool generates 15 waves per minute.
    The wavelength of these waves is $2 \cdot 0$ m.
    The frequency of the waves is

    A $0 \cdot 25$ Hz

    B $0 \cdot 50$ Hz

    C $4 \cdot 0$ Hz

    D   15 Hz

    E   30 Hz.

12. For a ray of light travelling from air into glass, which of the following statements is/are correct?

    I The speed of light always changes.

    II The speed of light sometimes changes.

    III The direction of light always changes.

    IV The direction of light sometimes changes.

    A   I only

    B   III only

    C   I and III only

    D   I and IV only

    E   II and IV

13. A ray of red light is incident on a glass block as shown.

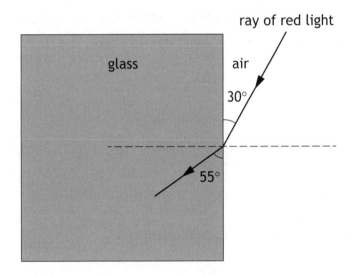

Which row in the table shows the values of the angle of incidence and angle of refraction?

	Angle of incidence	Angle of refraction
A	35°	60°
B	30°	55°
C	30°	35°
D	60°	55°
E	60°	35°

14. A student writes the following statements about the activity of a radioactive source.

    I The activity decreases with time.

    II The activity is measured in becquerels.

    III The activity is the number of decays per second.

Which of these statements is/are correct?

A   I only

B   II only

C   I and II only

D   II and III only

E   I, II and III

15. A worker in a nuclear power station is exposed to 3·0 mGy of gamma radiation and 0·50 mGy of fast neutrons.

The radiation weighting factor for gamma radiation is 1 and for fast neutrons is 10.

The total equivalent dose, in mSv, received by the worker is

A   3·50

B   8·00

C   30·5

D   35·0

E   38·5.

16. Which of the following contains two scalar quantities?

A   Force and mass

B   Weight and mass

C   Displacement and speed

D   Distance and speed

E   Displacement and velocity

17. A student sets up the apparatus as shown.

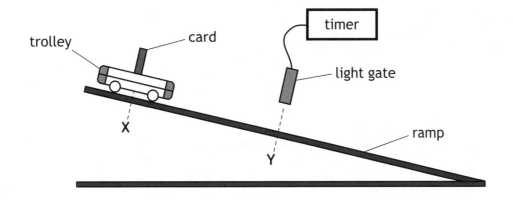

The trolley is released from X and moves down the ramp.

The following measurements are recorded.

time for card to pass through light gate = 0·08 s

distance from X to Y = 0·5 m

length of card = 40 mm

The instantaneous speed of the trolley at Y is

A   0·5 m s^{-1}

B   1·6 m s^{-1}

C   2·0 m s^{-1}

D   3·2 m s^{-1}

E   6·3 m s^{-1}.

18. As a car approaches a village the driver applies the brakes. The speed-time graph of the car's motion is shown.

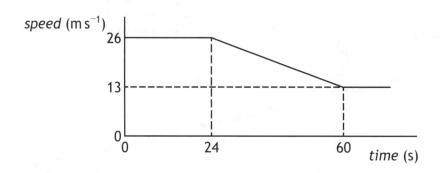

The brakes are applied for

A   13 s

B   20 s

C   24 s

D   36 s

E   60 s.

19. The Mars Curiosity Rover has a mass of 900 kg.

Which row of the table gives the mass and weight of the Rover on Mars?

	Mass (kg)	Weight (N)
A	243	243
B	243	900
C	900	900
D	900	3330
E	900	8820

20. An aircraft engine exerts a force on the air.

   Which of the following completes the "Newton pair" of forces?

   A   The force of the air on the aircraft engine

   B   The force of friction between the aircraft engine and the air

   C   The force of friction between the aircraft and the aircraft engine

   D   The force of the Earth on the aircraft engine

   E   The force of the aircraft engine on the Earth

**[END OF SECTION 1. NOW ATTEMPT THE QUESTIONS IN SECTION 2
OF YOUR QUESTION AND ANSWER BOOKLET]**

National Qualifications
SPECIMEN ONLY

SQ35/N5/11

Physics
Relationships Sheet

Date — Not applicable

$$E_p = mgh$$

$$E_k = \frac{1}{2}mv^2$$

$$Q = It$$

$$V = IR$$

$$R_T = R_1 + R_2 + \ldots$$

$$\frac{1}{R_T} = \frac{1}{R_1} + \frac{1}{R_2} + \ldots$$

$$V_2 = \left(\frac{R_2}{R_1 + R_2}\right)V_s$$

$$\frac{V_1}{V_2} = \frac{R_1}{R_2}$$

$$P = \frac{E}{t}$$

$$P = IV$$

$$P = I^2 R$$

$$P = \frac{V^2}{R}$$

$$E_h = cm\Delta T$$

$$p = \frac{F}{A}$$

$$\frac{pV}{T} = \text{constant}$$

$$p_1 V_1 = p_2 V_2$$

$$\frac{p_1}{T_1} = \frac{p_2}{T_2}$$

$$\frac{V_1}{T_1} = \frac{V_2}{T_2}$$

$$d = vt$$

$$v = f\lambda$$

$$T = \frac{1}{f}$$

$$A = \frac{N}{t}$$

$$D = \frac{E}{m}$$

$$H = Dw_R$$

$$\dot{H} = \frac{H}{t}$$

$$s = vt$$

$$d = \bar{v}t$$

$$s = \bar{v}t$$

$$a = \frac{v - u}{t}$$

$$W = mg$$

$$F = ma$$

$$E_w = Fd$$

$$E_h = ml$$

[END OF SPECIMEN RELATIONSHIPS SHEET]

N5

Mark

National Qualifications SPECIMEN ONLY

SQ35/N5/02

Physics Section 1— Answer Grid and Section 2

Date — Not applicable

Duration — 2 hours

Fill in these boxes and read what is printed below.

Full name of centre

Town

Forename(s)

Surname

Number of seat

Date of birth

Day	Month	Year
D D	M M	Y Y

Scottish candidate number

Total marks — 110

SECTION 1 — 20 marks

Attempt ALL questions in this section.

Instructions for completion of Section 1 are given on Page two.

SECTION 2 — 90 marks

Attempt ALL questions in this section.

Read all questions carefully before answering.

Use **blue** or **black** ink. Do NOT use gel pens.

Write your answers in the spaces provided. Additional space for answers and rough work is provided at the end of this booklet. If you use this space, write clearly the number of the question you are answering. Any rough work must be written in this booklet. You should score through your rough work when you have written your fair copy.

Before leaving the examination room you must give this booklet to the Invigilator. If you do not, you may lose all the marks for this paper.

SECTION 1 — 20 marks

The questions for Section 1 are contained in the booklet Physics Section 1 — Questions.
Read these and record your answers on the grid on Page three opposite.

1. The answer to each question is **either** A, B, C, D or E. Decide what your answer is, then fill in the appropriate bubble (see sample question below).

2. There is **only one correct** answer to each question.

3. Any rough working should be done on the rough working sheet.

Sample Question

The energy unit measured by the electricity meter in your home is the:

     A   ampere

     B   kilowatt-hour

     C   watt

     D   coulomb

     E   volt.

The correct answer is **B**—kilowatt-hour.   The answer **B** bubble has been clearly filled in (see below).

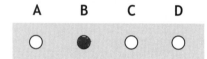

Changing an answer

If you decide to change your answer, cancel your first answer by putting a cross through it (see below) and fill in the answer you want. The answer below has been changed to **D**.

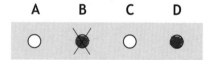

If you then decide to change back to an answer you have already scored out, put a tick (✓) to the **right** of the answer you want, as shown below:

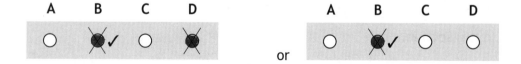

or

SECTION 1 — Answer Grid

	A	B	C	D	E
1	○	○	○	○	○
2	○	○	○	○	○
3	○	○	○	○	○
4	○	○	○	○	○
5	○	○	○	○	○
6	○	○	○	○	○
7	○	○	○	○	○
8	○	○	○	○	○
9	○	○	○	○	○
10	○	○	○	○	○
11	○	○	○	○	○
12	○	○	○	○	○
13	○	○	○	○	○
14	○	○	○	○	○
15	○	○	○	○	○
16	○	○	○	○	○
17	○	○	○	○	○
18	○	○	○	○	○
19	○	○	○	○	○
20	○	○	○	○	○

[BLANK PAGE]

SECTION 2 — 90 marks

Attempt ALL questions

MARKS | DO NOT WRITE IN THIS MARGIN

1. (a) A student sets up the following circuit.

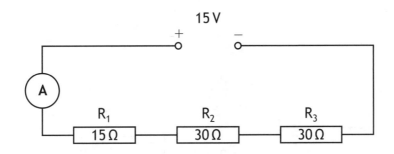

(i) Calculate the current in the circuit.

4

Space for working and answer

(ii) Calculate the potential difference across resistor R_1.

3

Space for working and answer

MARKS | DO NOT WRITE IN THIS MARGIN

1. **(continued)**

(b) The circuit is now rearranged as shown below.

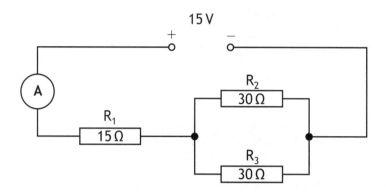

State how the reading on the ammeter compares to your answer in (a)(i).   **5**

Justify your answer by calculation.

Space for working and answer

Total marks 12

MARKS | DO NOT WRITE IN THIS MARGIN

2. A technician sets up a water bath for an experiment to study fermentation at different temperatures.

   The rating plate of the water bath is shown.

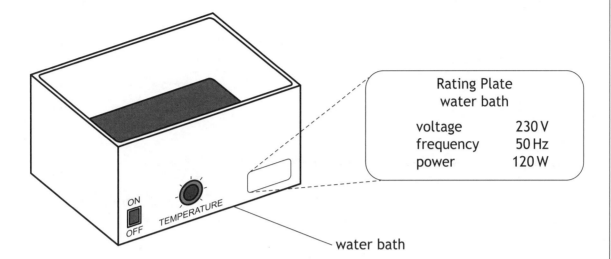

Rating Plate
water bath

voltage	230 V
frequency	50 Hz
power	120 W

   (a) The water bath contains 3·0 kg of water at an initial temperature of 15 °C.

   The specific heat capacity of the water is 4180 J kg^{-1} °C^{-1}.

   Calculate the energy required to raise the temperature of the water to 45 °C.   3

   Space for working and answer

   (b) Calculate the minimum time required to heat the water to 45 °C.   3

   Space for working and answer

MARKS | DO NOT WRITE IN THIS MARGIN

2. (continued)

(c) In practice it requires more time than calculated to heat the water.

(i) Explain why more time is required.   1

(ii) Suggest one way of reducing this additional time.   1

Total marks   8

2. (continued)

MARKS DO NOT WRITE IN THIS MARGIN

3. Extreme temperatures have been known to cause some electricity supply pylons to collapse.

Using your knowledge of physics, comment on why this happens.   3

4. Architects need to know how well different materials insulate buildings. This can be determined using U-values.

The U-value is defined as the rate at which heat energy is transferred through one square metre of building material when the temperature difference is one degree Celsius.

The rate of heat transfer through a material can be determined using:

rate of heat transfer = U-value × area × difference in temperature

The tables below give information for two houses.

House P

House P	U-value ($W\,m^{-2}\,°C^{-1}$)	Total area (m^2)
Uninsulated roof	2·0	150
Cavity walls	1·9	300
Single glazed windows	5·6	50

House Q

House Q	U-value ($W\,m^{-2}\,°C^{-1}$)	Total area (m^2)
Insulated roof	0·5	150
Filled cavity walls	0·6	500
Double glazed windows	2·8	80

MARKS

4. (continued)

(a) Complete the sentence below by circling the correct answer.   1

The $\begin{Bmatrix} \text{higher} \\ \text{lower} \end{Bmatrix}$ the U-value, the better the material is as a heat

insulator.

(b) Show by calculation that house P has the highest rate of heat transfer through the **walls** when the outside temperature is 2 °C and the inside temperature in both houses is 18 °C.   4

Space for working and answer

MARKS

4.　(continued)

(c)　Glass transmits infrared radiation and visible light.　The percentage transmitted depends on the type and thickness of the glass.　The data from tests on two different types of glass is displayed in the graph below.

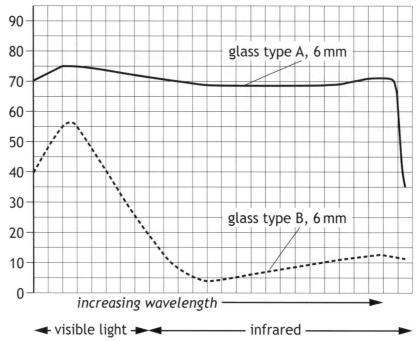

A glass conservatory is being built on house Q. The homeowner wants the inside of the conservatory to remain as cool as possible throughout the summer.

Using information from the graph, explain which type of glass should be used.

2

Total marks　7

MARKS | DO NOT WRITE IN THIS MARGIN

5. A pair of neutron stars which orbit one another will over time move closer together and eventually join.

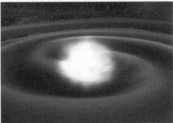

Astronomers believe that as the neutron stars move closer, they emit energy in the form of gravitational waves. It is predicted that gravitational wave detectors will produce the graphs shown.

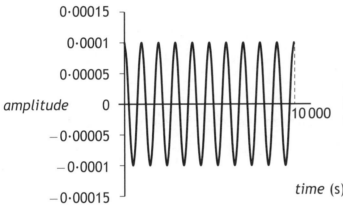

One million years before stars join together

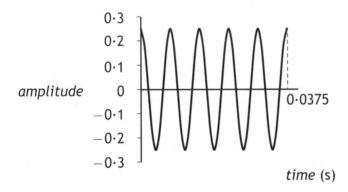

One second before stars join together

0·1 seconds before stars join together

5. **(continued)**

(a) Use the graphs to complete the following table. The first row has already been completed.

4

Time before the stars join	Period of gravitational waves (s)	Frequency of gravitational waves (Hz)
1 million years	1000	0·001
1 second		
0·1 second		

Space for working

(b) State what happens to the frequency of the gravitational waves as the neutron stars move closer together.

1

(c) The orbital speed, in metres per second, of the rotating neutron stars is given by the equation:

$$v = \frac{2\pi}{T}R$$

where T is the orbital period in seconds and R is half the distance between the stars in metres.

Calculate the orbital speed of the neutron stars when they are 340 000 km apart and the orbital period is 1150 s.

2

Space for working and answer

Total marks   **7**

MARKS | DO NOT WRITE IN THIS MARGIN

6. A water wave is diffracted when it passes through a gap in a barrier. The wavelength of the wave is 10 mm. The gap is less than 10 mm.

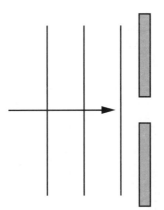

(a) Complete the diagram above to show the pattern of the wave to the right of the barrier.

2

(b) The diagram below represents the electromagnetic spectrum.

Radio & TV waves	A	Infrared radiation	Visible light	Ultraviolet light	X-rays	Gamma radiation

(i) Identify radiation A.

1

(ii) Apart from diffraction, state one property that all electromagnetic waves have in common.

1

Total marks   4

MARKS | DO NOT WRITE IN THIS MARGIN

7. Trees continually absorb carbon-14 when they are alive. When a tree dies the carbon-14 contained in its wood is not replaced. Carbon-14 is radioactive and decays by beta emission.

(a) Following the tree's death, the activity of the carbon-14 within a **25 mg** sample of its wood changes as shown.

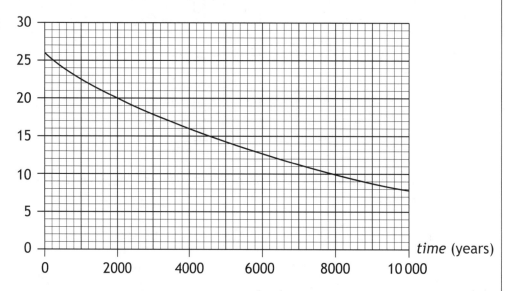

(i) Use the graph to determine the half-life of carbon-14.   **2**

(ii) Calculate the time taken for the activity of this sample of carbon-14 to fall to 6·5 Bq.   **3**

Space for working and answer

MARKS | DO NOT WRITE IN THIS MARGIN

7. (a) (continued)

(iii) During an archaeological dig, a 125 mg sample of the same type of wood was obtained. The activity of this sample was 40 Bq.

Estimate the age of this sample.   3

Space for working and answer

(b) Explain why this method could not be used to estimate the age of a tree that died 100 years ago.   1

Total marks   9

MARKS | DO NOT WRITE IN THIS MARGIN

8. A technician uses a radioactive source to investigate the effect of gamma rays on biological tissue.

(a) State what is meant by the term *gamma rays*.　　　1

(b) The wavelength of a gamma ray is $6{\cdot}0 \times 10^{-13}$ m.

Calculate the frequency of the gamma ray.　　　3

Space for working and answer

(c) In one experiment, a biological tissue sample of mass $0{\cdot}10$ kg receives an absorbed dose of $50\,\mu$Gy.

Calculate the energy absorbed by the tissue.　　　3

Space for working and answer

MARKS | DO NOT WRITE IN THIS MARGIN

8. (continued)

(d) The radioactive source must be stored in a lead-lined container.

Explain why a lead-lined container should be used.   1

Total marks   8

MARKS | DO NOT WRITE IN THIS MARGIN

9. An aircraft is making a journey between two airports. A graph of the aircraft's velocity during take-off is shown below.

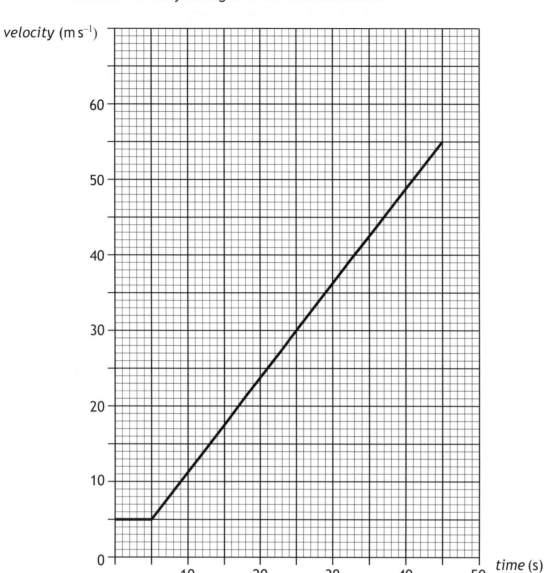

(a) Calculate the acceleration during take-off.   3

Space for working and answer

MARKS | DO NOT WRITE IN THIS MARGIN

9.   (continued)

(b)   (i)   During flight, the aircraft is travelling at a velocity of $150\,m\,s^{-1}$ due north and then encounters a crosswind of $40\,m\,s^{-1}$ due east.

By scale diagram, or otherwise, determine the resultant velocity of the aircraft.

4

N

$150\,ms^{-1}$

$40\,m\,s^{-1}$

Space for working and answer

MARKS | DO NOT WRITE IN THIS MARGIN

9. (b) (continued)

(ii) Describe what action the pilot could take to ensure that the aircraft remains travelling north at 150 m s⁻¹. **2**

(c) The aircraft arrives at the destination airport.

This airport has three runways of different lengths to accommodate different sizes of aircraft.

Explain why larger aircraft require a much longer runway to land safely. **2**

Total marks 11

MARKS

10. The Soyuz Spacecraft is used to transport astronauts to the International Space Station (ISS). The spacecraft contains three parts that are launched together.

Part	Mass (kg)
Orbital Module	1300
Descent Module (including astronauts)	2950
Instrumentation/ Propulsion Module	2900

(a) When the spacecraft leaves the ISS, its propulsion module produces a force of 1430 kN.

Calculate the acceleration of the spacecraft as it leaves the ISS.

Space for working and answer

4

MARKS | DO NOT WRITE IN THIS MARGIN

10. **(continued)**

(b) On the return flight, the Orbital Module and the Instrumentation/Propulsion Module are jettisoned. Instead of returning to Earth, they burn up in the atmosphere at a very high temperature.

Explain why these Modules burn up on re-entry into the atmosphere.　　　　**2**

(c) After the Descent Module has re-entered the atmosphere, its speed is dramatically reduced.

　　(i) Four parachutes are used to slow the Module's rate of descent from $230\,\text{m s}^{-1}$ to $80\,\text{m s}^{-1}$.

　　　Explain, in terms of forces, how the parachutes reduce the speed of the Module.　　　　**2**

MARKS | DO NOT WRITE IN THIS MARGIN

10. **(c)** **(continued)**

(ii) Just before touchdown, small engines fire on the bottom of the Module, slowing it down further. The work done by the engines is 80 kJ over a distance of 5 m.

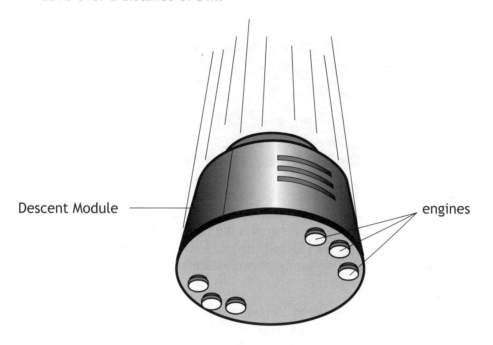

Descent Module ⎯⎯⎯   engines

Calculate the force produced by the engines.   **3**

Space for working and answer

Total marks 11

MARKS | DO NOT WRITE IN THIS MARGIN

11. Read the passage below and answer the questions that follow.

Dragonfish nebula conceals giant cluster of young stars

The Dragonfish nebula may contain the Milky Way's most massive cluster of young stars. Scientists from the University of Toronto found the first hint of the cluster in 2010 in the form of a big cloud of ionised gas 30 000 light years from Earth. They detected the gas from its microwave emissions, suspecting that radiation from massive stars nearby had ionised the gas.

Now the scientists have identified a cluster of 400 massive stars in the heart of the gas cloud using images from an infrared telescope. The cluster probably contains more stars which are too small and dim to detect.

The surrounding cloud of ionised gas is producing more microwaves than the clouds around other star clusters in our galaxy. This suggests that the Dragonfish nebula contains the brightest and most massive young cluster discovered so far, with a total mass of around 100 000 times the mass of the Sun.

(a) Name the galaxy mentioned in the passage.

1

(b) Show that the Dragonfish nebula is approximately $2·84 \times 10^{20}$ m away from Earth.

3

Space for working and answer

MARKS | DO NOT WRITE IN THIS MARGIN

11. **(continued)**

(c) Complete the sentence by circling the correct words.

Compared to infrared radiation, microwaves have a $\left\{\begin{array}{c}\text{longer}\\\text{shorter}\end{array}\right\}$ wavelength which means they have a $\left\{\begin{array}{c}\text{higher}\\\text{lower}\end{array}\right\}$ frequency.

1

(d) A line spectrum from a nebula is shown below.

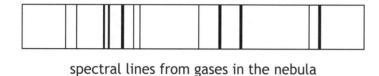

spectral lines from gases in the nebula

nitrogen

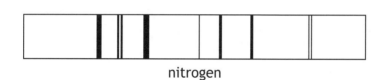

helium

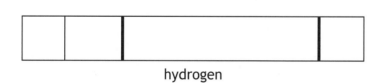

hydrogen

krypton

Identify the elements present in the nebula.

2

Total marks 7

MARKS

DO NOT WRITE IN THIS MARGIN

12. In October 2012, a skydiver jumped from a balloon at a height of 39 km above the surface of the Earth.

He became the first person to jump from this height.

He also became the first human to fall at speeds higher than the speed of sound in air.

Using your knowledge of physics, comment on the challenges faced by the skydiver when making this jump.

3

Space for answer

[END OF SPECIMEN QUESTION PAPER]

MARKS

DO NOT WRITE IN THIS MARGIN

ADDITIONAL SPACE FOR ROUGH WORKING AND ANSWERS

MARKS DO NOT WRITE IN THIS MARGIN

ADDITIONAL SPACE FOR ROUGH WORKING AND ANSWERS

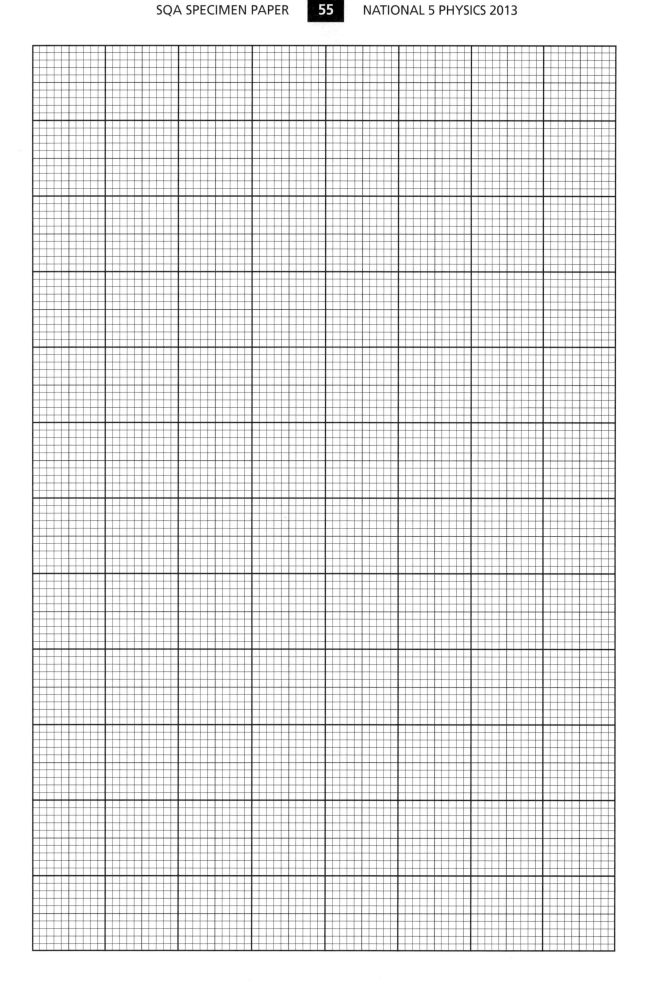

Acknowledgement of Copyright

Section 2 Question 11 Extract is adapted from an article titled "Dragonfish nebula conceals giant star cluster" taken from the New Scientist Magazine, 26 January 2011. Reproduced by kind permission of New Scientist.

2013 Model Paper 1

HODDER
GIBSON
LEARN MORE

National Qualifications
MODEL PAPER 1

Physics
Section 1—Questions

Date — Not applicable

Duration — 2 hours

Instructions for completion of Section 1 are given on Page two of the question paper.

Record your answers on the grid on Page three of your answer booklet

Do NOT write in this booklet.

Before leaving the examination room you must give your answer booklet to the Invigilator. If you do not, you may lose ALL the marks for this paper.

HODDER
GIBSON
LEARN MORE

DATA SHEET

Speed of light in materials

Material	Speed in m s^{-1}
Air	$3 \cdot 0 \times 10^8$
Carbon dioxide	$3 \cdot 0 \times 10^8$
Diamond	$1 \cdot 2 \times 10^8$
Glass	$2 \cdot 0 \times 10^8$
Glycerol	$2 \cdot 1 \times 10^8$
Water	$2 \cdot 3 \times 10^8$

Gravitational field strengths

	Gravitational field strength on the surface in N kg^{-1}
Earth	$9 \cdot 8$
Jupiter	23
Mars	$3 \cdot 7$
Mercury	$3 \cdot 7$
Moon	$1 \cdot 6$
Neptune	11
Saturn	$9 \cdot 0$
Sun	270
Uranus	$8 \cdot 7$
Venus	$8 \cdot 9$

Specific latent heat of fusion of materials

Material	Specific latent heat of fusion in J kg^{-1}
Alcohol	$0 \cdot 99 \times 10^5$
Aluminium	$3 \cdot 95 \times 10^5$
Carbon Dioxide	$1 \cdot 80 \times 10^5$
Copper	$2 \cdot 05 \times 10^5$
Iron	$2 \cdot 67 \times 10^5$
Lead	$0 \cdot 25 \times 10^5$
Water	$3 \cdot 34 \times 10^5$

Specific latent heat of vaporisation of materials

Material	Specific latent heat of vaporisation in J kg^{-1}
Alcohol	$11 \cdot 2 \times 10^5$
Carbon Dioxide	$3 \cdot 77 \times 10^5$
Glycerol	$8 \cdot 30 \times 10^5$
Turpentine	$2 \cdot 90 \times 10^5$
Water	$22 \cdot 6 \times 10^5$

Speed of sound in materials

Material	Speed in m s^{-1}
Aluminium	5200
Air	340
Bone	4100
Carbon dioxide	270
Glycerol	1900
Muscle	1600
Steel	5200
Tissue	1500
Water	1500

Specific heat capacity of materials

Material	Specific heat capacity in J kg^{-1} °C^{-1}
Alcohol	2350
Aluminium	902
Copper	386
Glass	500
Ice	2100
Iron	480
Lead	128
Oil	2130
Water	4180

Melting and boiling points of materials

Material	Melting point in °C	Boiling point in °C
Alcohol	−98	65
Aluminium	660	2470
Copper	1077	2567
Glycerol	18	290
Lead	328	1737
Iron	1537	2737

Radiation weighting factors

Type of radiation	Radiation weighting factor
alpha	20
beta	1
fast neutrons	10
gamma	1
slow neutrons	3

SECTION 1

1. In the circuit shown below, R_1 and R_2 are two identical resistors connected in series across a 24 V supply.

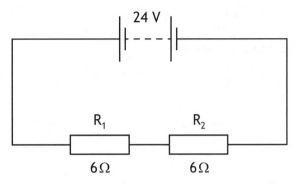

What is the current in resistor R_1?

A   0·25 A

B   0·5 A

C   2 A

D   4 A

E   8 A

2. Three resistors are connected as shown:

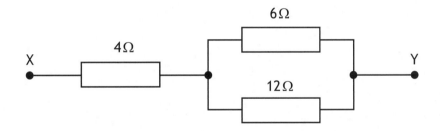

The total resistance between X and Y is

A   2Ω

B   4Ω

C   8Ω

D   13Ω

E   22Ω.

3. The resistance of a wire is 6 Ω.

   The current in the wire is 2A.

   The power developed in the wire is

   A   2W

   B   3W

   C   18W

   D   24W

   E   72W.

4. Which of the following is the correct symbol for a transistor?

   A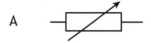

   B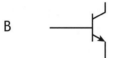

   C

   D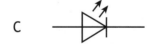

   E

5. A student sets up the circuits shown.

   In which circuit will both LEDs be lit?

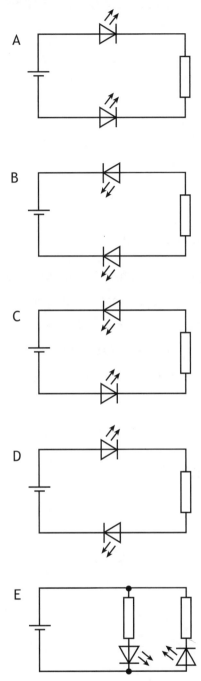

6. A passenger aircraft cruises at an altitude where the outside air pressure is $0{\cdot}35{\times}10^5$ Pa.

   The air pressure inside the aircraft is $1{\cdot}0{\times}10^5$ Pa. The area of an external cabin door is $2{\cdot}2$ m^2.

   What is the outward force on the door due to the pressure difference?

   A $0{\cdot}30{\times}10^5$ N

   B $0{\cdot}61{\times}10^5$ N

   C $1{\cdot}43{\times}10^5$ N

   D $2{\cdot}2{\times}10^5$ N

   E $2{\cdot}97{\times}10^5$ N.

7. For a fixed mass of gas at constant volume

   A   the pressure is directly proportional to temperature in K

   B   the pressure is inversely proportional to temperature in K

   C   the pressure is directly proportional to temperature in °C

   D   the pressure is inversely proportional to temperature in °C

   E   (pressure × temperature in K) is constant.

8. Ice at a temperature of −10 °C is heated until it becomes water at 80 °C.

   The temperature change in kelvin is

   A   70 K

   B   90 K

   C   343 K

   D   363 K

   E   636 K.

9. The energy of a water wave depends on its

   A   wavelength

   B   period

   C   colour

   D   speed

   E   amplitude.

10. A student can hear sound ranging from 20 Hz to 20 kHz.

If the velocity of sound in air is 340 m s^{-1}, the shortest wavelength the student can hear is

A   0·017 m

B   0·17 m

C   1·7 m

D   17 m

E   170 m.

11. The following diagram gives information about a wave.

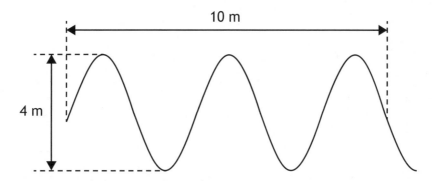

Which row shows the amplitude and wavelength of the wave?

	Amplitude (m)	Wavelength (m)
A	2	2
B	2	4
C	2	5
D	4	2
E	4	4

12. Which of the following lists the particles in order of size from smallest to largest?

    A   helium nucleus; electron; proton

    B   helium nucleus; proton; electron

    C   proton; helium nucleus; electron

    D   electron; helium nucleus; proton

    E   electron; proton; helium nucleus

13. A light wave has a frequency of $5 \cdot 2 \times 10^{14}$ Hz.

The period of the wave is

    A $1 \cdot 9 \times 10^{-15}$ s

    B $5 \cdot 8 \times 10^{-7}$ s

    C $0 \cdot 19$ s

    D $1 \cdot 7 \times 10^{6}$ s

    E $1 \cdot 9 \times 10^{15}$ s.

14. A student makes the following statements for a ray of light travelling from glass into air.

    I   The direction of light always changes.

    II   The direction of light sometimes changes.

    III   The speed of light always changes.

    IV   The speed of light sometimes changes.

Which of these statements is/are correct?

    A   I and III only

    B   II and III only

    C   I and IV only

    D   III only

    E   IV only

15. A student makes the following statements about ionising radiations.

   I   Ionisation occurs when an atom loses an electron.

   II   Gamma radiation produces greater ionisation (density) than alpha particles.

   III   An alpha particle consists of 2 protons, 2 neutrons and 2 electrons.

   Which of these statements is/are correct?

   A   I only

   B   II only

   C   I and II only

   D   II and III only

   E   I, II and III

16. Which of the following contains two scalar quantities?

   A   Weight and mass

   B   Distance and speed

   C   Force and mass

   D   Displacement and velocity

   E   Force and mass

17. The graph shows how the velocity of a ball changes with time.

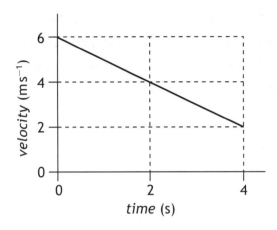

The acceleration of the ball is

A $-8\,\mathrm{ms^{-2}}$

B $8\,\mathrm{ms^{-2}}$

C $24\,\mathrm{ms^{-2}}$

D $-1\,\mathrm{ms^{-2}}$

E $1\,\mathrm{ms^{-2}}$.

18. The Apollo Lunar Rover vehicle used by astronauts on the Moon had a mass of 204 kg.

Which row of the table gives the mass and weight of the vehicle on the moon?

	Mass (kg)	Weight (N)
A	55	55
B	55	204
C	204	204
D	204	326
E	204	1999

19. A ball is thrown horizontally from a cliff as shown.

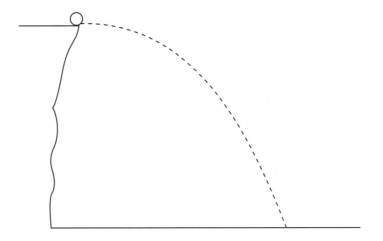

The effect of air resistance is negligible.

A student makes the following statements about the ball.

I   The vertical speed of the ball increases as it falls.

II   The vertical acceleration of the ball increases as it falls.

III The vertical force on the ball increases as it falls.

Which of these statements is/are correct?

A   I only

B   II only

C   I and II only

D   II and III only

E   I, II and III

20. A rocket of mass 200 kg accelerates vertically upwards from the surface of a planet at $2 \cdot 0$ m s^{-2}.

The gravitational field strength on the planet is $4 \cdot 0$ N kg^{-1}.

What is the size of the force being exerted by the rocket's engines?

A   400 N

B   800 N

C   1200 N

D   2000 N

E   2400 N

**[END OF SECTION 1. NOW ATTEMPT THE QUESTIONS IN SECTION 2
OF YOUR QUESTION AND ANSWER BOOKLET]**

National Qualifications
MODEL PAPER 1

Physics
Relationships Sheet

Date — Not applicable

$$E_p = mgh$$

$$E_k = \tfrac{1}{2}mv^2$$

$$Q = It$$

$$V = IR$$

$$R_T = R_1 + R_2 + \ldots$$

$$\frac{1}{R_T} = \frac{1}{R_1} + \frac{1}{R_2} + \ldots$$

$$V_2 = \left(\frac{R_2}{R_1 + R_2}\right)V_s$$

$$\frac{V_1}{V_2} = \frac{R_1}{R_2}$$

$$P = \frac{E}{t}$$

$$P = IV$$

$$P = I^2 R$$

$$P = \frac{V^2}{R}$$

$$E_h = cm\Delta T$$

$$p = \frac{F}{A}$$

$$\frac{pV}{T} = \text{constant}$$

$$p_1 V_1 = p_2 V_2$$

$$\frac{p_1}{T_1} = \frac{p_2}{T_2}$$

$$\frac{V_1}{T_1} = \frac{V_2}{T_2}$$

$$d = vt$$

$$v = f\lambda$$

$$T = \frac{1}{f}$$

$$A = \frac{N}{t}$$

$$D = \frac{E}{m}$$

$$H = Dw_R$$

$$\dot{H} = \frac{H}{t}$$

$$s = vt$$

$$d = \bar{v}t$$

$$s = \bar{v}t$$

$$a = \frac{v - u}{t}$$

$$W = mg$$

$$F = ma$$

$$E_w = Fd$$

$$E_h = ml$$

[END OF SPECIMEN RELATIONSHIPS SHEET]

Page two

N5 National Qualifications MODEL PAPER 1

Physics Section 1—
Answer Grid and
Section 2

Duration — 2 hours

Total marks — 110

SECTION 1 — 20 marks

Attempt ALL questions in this section.

Instructions for completion of Section 1 are given on Page two.

SECTION 2 — 90 marks

Attempt ALL questions in this section.

Read all questions carefully before answering.

Use **blue** or **black** ink. Do NOT use gel pens.

Write your answers in the spaces provided. Additional space for answers and rough work is provided at the end of this booklet. If you use this space, write clearly the number of the question you are answering. Any rough work must be written in this booklet. You should score through your rough work when you have written your fair copy.

Before leaving the examination room you must give this booklet to the Invigilator. If you do not, you may lose all the marks for this paper.

SECTION 1 — 20 marks

The questions for Section 1 are contained in the booklet Physics Section 1 — Questions.
Read these and record your answers on the grid on Page three opposite.

1. The answer to each question is **either** A, B, C, D or E. Decide what your answer is, then fill in the appropriate bubble (see sample question below).

2. There is **only one correct** answer to each question.

3. Any rough working should be done on the rough working sheet.

Sample Question

The energy unit measured by the electricity meter in your home is the:

     A   ampere

     B   kilowatt-hour

     C   watt

     D   coulomb

     E   volt.

The correct answer is **B**—kilowatt-hour. The answer **B** bubble has been clearly filled in (see below).

Changing an answer

If you decide to change your answer, cancel your first answer by putting a cross through it (see below) and fill in the answer you want. The answer below has been changed to **D**.

If you then decide to change back to an answer you have already scored out, put a tick (✓) to the **right** of the answer you want, as shown below:

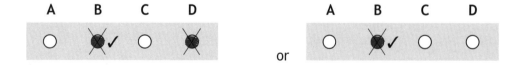

                               or

SECTION 1 — Answer Grid

	A	B	C	D	E
1	○	○	○	○	○
2	○	○	○	○	○
3	○	○	○	○	○
4	○	○	○	○	○
5	○	○	○	○	○
6	○	○	○	○	○
7	○	○	○	○	○
8	○	○	○	○	○
9	○	○	○	○	○
10	○	○	○	○	○
11	○	○	○	○	○
12	○	○	○	○	○
13	○	○	○	○	○
14	○	○	○	○	○
15	○	○	○	○	○
16	○	○	○	○	○
17	○	○	○	○	○
18	○	○	○	○	○
19	○	○	○	○	○
20	○	○	○	○	○

[BLANK PAGE]

SECTION 2 — 90 marks

Attempt ALL questions

1. While repairing a school roof, workmen lift a pallet of tiles from the ground to the top of the scaffolding.

   This job is carried out using a motorised pulley system.

   The pallet and tiles have a total mass of 235 kg.

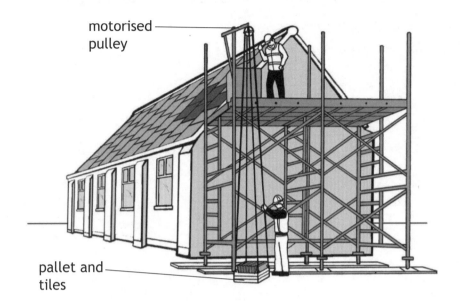

motorised pulley

pallet and tiles

   (a) The pallet and tiles are lifted to a height of 12 m.

   Calculate the gravitational potential energy gained by the pallet and tiles.

   Space for working and answer.

   3

$$E_p = mgh$$
$$E_p = 235 \times 9.8 \times 12$$
$$E_p = 27636 \ J$$

MARKS | DO NOT WRITE IN THIS MARGIN

1. (continued)

(b) When the tiles are being unloaded onto the scaffolding, at a height of 12 m, one tile falls.

The tile has a mass of 2·5 kg.

(i) Calculate the final speed of the tile just before it hits the ground.

Assume the tile falls from rest.   4

Space for working and answer.

$$E_p = mgh \qquad E_K = \tfrac{1}{2}mv^2$$
$$E_p = 2.5 \times 9.8 \times 12 \qquad 294 = 0.5 \times 2.5 \times v^2$$
$$E_p = 294 \text{ J} \qquad v = \sqrt{\frac{294}{0.5 \times 2.5}}$$
$$v = 15.3362316$$
$$v = 15.3 \text{ ms}^{-1}$$

(ii) Explain why the actual speed is less than the speed calculated in (b)(i).   1

The tile is slowed down by air resistance

Total marks   8

MARKS

2. A student sets up the following circuit to investigate the resistance of resistor R.

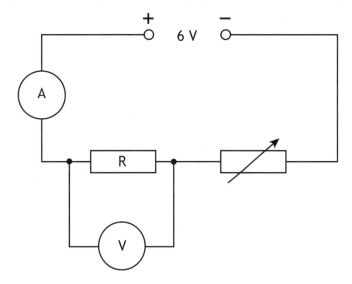

The variable resistor is adjusted and the voltmeter and ammeter readings are noted. The following graph is obtained from the experimental results.

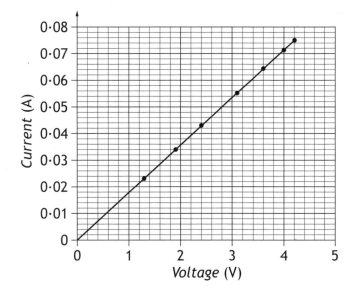

(a) (i) Calculate the value of the resistor R when the reading on the voltmeter is 4·2 V.   3

Space for working and answer.

$$R = \frac{V}{I}$$

$$R = \frac{4.2}{0.075}$$

$$R = 56 \ \Omega$$

MARKS

2.　(continued)

(ii)　Using information from the graph, state whether the resistance of the resistor R, increases, stays the same or decreases as the voltage increases.

You must justify your answer.　　**2**

increases - the graph is a straight line

(b)　The student is given a task to combine two resistors from a pack containing one each of 33 Ω, 56 Ω, 82 Ω, 150 Ω, 270 Ω, 390 Ω.

Show by calculation which two resistors should be used to give the smallest combined resistance.　　**3**

Space for working and answer.

Total marks　8

MARKS

3. An important experiment which resulted in determining the value for the charge on an electron was carried out by a scientist named Millikan in 1909.

   Part of the experiment required the **terminal velocity** of tiny drops of oil to be determined as they fell through air.

   (a) Explain what is meant by terminal velocity.   **1**

   (b) As the oil drop fell through the air, Milliken used an equation known as Stokes' Law to determine the upward drag force, F_d, acting on the drop:

   $$F_d = 6\pi r \eta v_1$$

   Where:

   v_1 is the terminal velocity of the falling drop,

   η is the viscosity of the air,

   r is the radius of the drop

   For one particular oil drop, its radius was $2 \cdot 83$ μm, its terminal velocity was $8 \cdot 56 \times 10^{-4}$ m s^{-1} and the viscosity of air was $1 \cdot 820 \times 10^{-5}$ kg m^{-1} s^{-1}.

   Calculate the value of the drag force, F_d, acting on the oil drop.   **2**

   Space for working and answer.

MARKS | DO NOT WRITE IN THIS MARGIN

3. **(continued)**

(c) The weight of another oil drop was $8 \cdot 6 \times 10^{-13}$ N.

Calculate the mass of this oil drop.   **3**
Space for working and answer.

(d) For another part of the experiment, a quantity of charge was added to the oil drop. The drop was then placed in an electric field.

State the effect of an electric field on a charged particle.   **1**

Total marks   7

MARKS | DO NOT WRITE IN THIS MARGIN

4. A solar furnace consists of an array of mirrors which reflect heat radiation on to a central curved reflector.

A heating container is placed at the focus of the central curved reflector.

Metals placed in the container are heated until they melt.

(a) 8000 kg of pre-heated aluminium pellets at a temperature of 160°C are placed in the container. Aluminium has a specific heat capacity of 902 J kg^{-1}°C^{-1} and a melting point of 660°C.

Calculate the heat energy required to heat the aluminium to its melting point.   **3**

Space for working and answer.

(b) How much extra energy is required to melt the aluminium pellets?   **3**
Space for working and answer.

Total marks   **6**

5. Estimate the pressure exerted on the floor by an average National 5 student who is standing on two feet.

   Show any working clearly, and explain any assumptions that you make.　　3

MARKS

6. A student is training to become a diver.

The student carries out an experiment to investigate the relationship between the pressure and volume of a fixed mass of gas using the apparatus shown.

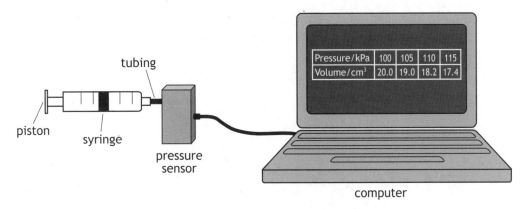

Pressure/kPa	100	105	110	115
Volume/cm³	20.0	19.0	18.2	17.4

tubing

piston

syringe

pressure sensor

computer

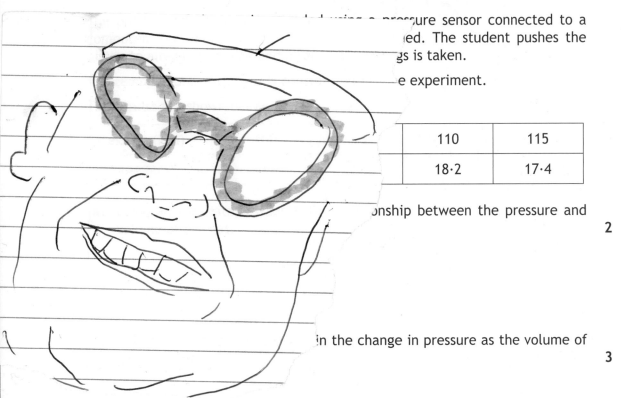

...ure sensor connected to a ...ed. The student pushes the ...gs is taken.

...e experiment.

	110	115
	18·2	17·4

...onship between the pressure and

2

...in the change in pressure as the volume of

3

(c) Explain why it is important for the tubing to be as short as possible. 1

Total marks 6

MARKS

7. (a) Refraction of light occurs in glass.

What is meant by the term refraction?   **1**

(b) The following diagram shows a ray of light entering a glass block.

(i) Complete the diagram to show the path of the ray of light through the block and after it emerges from the block.   **2**

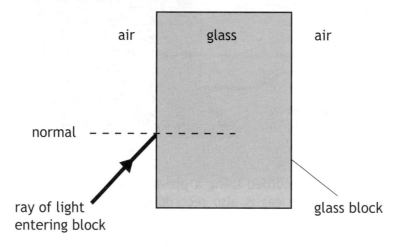

(ii) On your diagram indicate an angle of refraction, r.   **1**

(c) The Sun produces electromagnetic radiation. The electromagnetic spectrum is shown in order of increasing wavelength. Two radiations P and Q have been omitted.

Gamma rays	X rays	P	Visible light	Infra red	Q	Television and radio waves

increasing wavelength

MARKS | DO NOT WRITE IN THIS MARGIN

7. **(continued)**

    (i) Identify radiations P and Q.   **2**

    (ii) The planet Neptune is $4 \cdot 50 \times 10^9$ km from the Sun.

        Calculate the time taken for radio waves from the Sun to reach Neptune.   **3**

        Space for working and answer.

Total marks **9**

8. In 1908 Ernest Rutherford conducted a series of experiments involving alpha particles.

(a) State what is meant by an alpha particle.   1

(b) Alpha particles produce a greater ionisation density than beta particles or gamma rays.

What is meant by the term *ionisation*?   1

MARKS | DO NOT WRITE IN THIS MARGIN

8. **(continued)**

(c) A radioactive source emits alpha particles and has a half-life of 2·5 hours.

The source has an initial activity of 4·8 kBq.

Calculate the time taken for its activity to decrease to 300 Bq.   3

Space for working and answer.

(d) Some sources emit alpha particles and are stored in lead cases despite the fact that alpha particles cannot penetrate paper.

Suggest a possible reason for storing these sources using this method.   1

Total marks   6

9. An ageing nuclear power station is being dismantled.

(a) During the dismantling process a worker comes into contact with an object that emits 24 000 alpha particles in five minutes. The worker's hand has a mass of 0·50 kg and absorbs 6·0 µJ of energy.

   (i) Calculate the absorbed dose received by the worker's hand.   **3**

     Space for working and answer.

MARKS | DO NOT WRITE IN THIS MARGIN

9. (continued)

   (ii) Calculate the equivalent dose received by the worker's hand.   **3**
   Space for working and answer.

   (iii) Calculate the activity of the object.   **3**
   Space for working and answer.

(b) What type of nuclear reaction takes place in a nuclear power station's reactor?   **1**

Total marks 10

10. Two cyclists choose different routes to travel from point **A** to a point **B** some distance away.

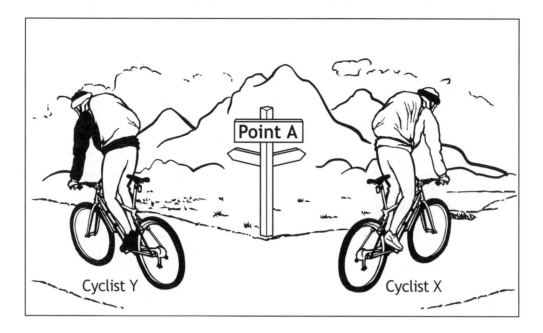

(a) Cyclist X travels 12 km due East (090).

He then turns and travels due South (180) and travels a further 15 km to arrive at **B**.

He takes 1 hour 15 minutes to travel from **A** to **B**.

(i) By scale drawing (or otherwise) find the displacement of **B** from **A**.   4

MARKS | DO NOT WRITE IN THIS MARGIN

10. (continued)

(ii) Calculate the average velocity of cyclist X for the journey from **A** to **B**. 3

Space for working and answer.

(b) Cyclist Y travels a total distance of 33 km by following a different route from **A** to **B** at an average speed of 22 $km\,h^{-1}$.

State the displacement of cyclist Y on completing this route. 1

Total marks 8

MARKS | DO NOT WRITE IN THIS MARGIN

11. A child sledges down a hill.

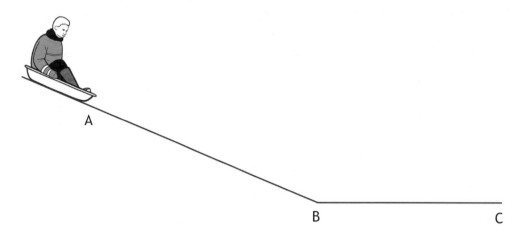

The sledge and child are released from rest at point A. They reach a speed of $3\,m\,s^{-1}$ at point B.

(a) The sledge and child take 5s to reach point B.

Calculate the acceleration.   **3**

Space for working and answer.

(b) The sledge and child have a combined mass of 40 kg.

Calculate the unbalanced force acting on them.   **3**

Space for working and answer.

Total marks   **6**

MARKS | DO NOT WRITE IN THIS MARGIN

12. An underwater generator is designed to produce electricity from water currents in the sea.

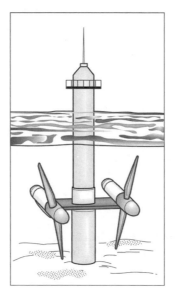

The output power of the generator depends on the speed of the water current as shown in Graph 1.

Graph 1

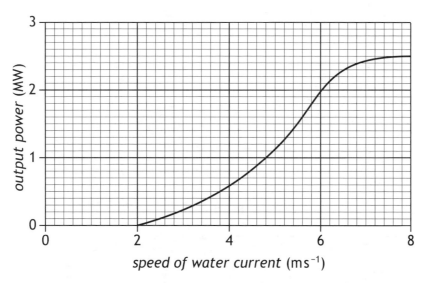

The speed of the water current is recorded at different times of the day shown in Graph 2.

Graph 2

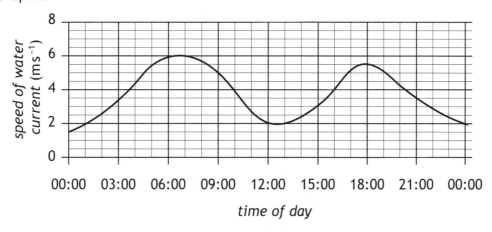

MARKS

12. (continued)

(a) (i) State the output power of the generator at 09:00.   1

(ii) State one disadvantage of using this type of generator.   1

(b) Three different types of electrical generator, X, Y and Z are tested in a special tank with a current of water as shown to find out the efficiency of each generator.

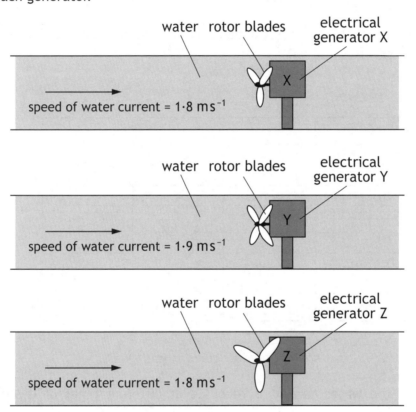

Give **two** reasons why this is not a fair test.   2

Total marks   4

13. Read the passage below and answer the questions that follow.

Black Holes

A black hole is where the force of gravity is so strong because matter has collapsed into a tiny space. This can happen when a star has used up most of its energy.

Because no light can get out, black holes cannot be seen. Space telescopes can help find black holes. Using special instruments on satellites, the behaviour of stars which are close to black holes can be monitored.

Black holes can have a range of sizes. A 'stellar' black hole has a mass of up to 20 times more than the mass of the sun. There may be many stellar mass black holes in the Milky Way galaxy.

There is evidence which suggests that at the centre of the Milky Way there is a supermassive black hole.

A supermassive black hole has a mass of more than 1 million suns. Scientists have discovered that every large galaxy contains a supermassive black hole at its centre.

At a distance of $2 \cdot 6 \times 10^{20}$ metres from Earth, the supermassive black hole at the centre of the Milky Way is called Sagittarius A. Sagittarius A would fit inside a large sphere which could hold several million Earth masses.

Scientists think the supermassive black holes formed when the universe began.

Stellar black holes are made when the centre of a very big star collapses inwards on itself. When this happens, it causes a supernova. A supernova is an exploding star that blasts part of the star into space.

Black holes cannot be seen directly because gravity prevents light escaping from the black hole. Observation of how nearby stars are affected by their strong gravity provides information about the behaviour, size and nature of the black hole.

The interaction of stars and black holes when they are close together produces intense gamma radiation. Satellites and telescopes in space are used to detect this radiation.

MARKS | DO NOT WRITE IN THIS MARGIN

13. **(continued)**

(a) Name a black hole mentioned in the passage.   1

(b) Calculate how many light years the Earth is from the centre of the Milky Way.   3

Space for working and answer.

(c) Telescopes on satellites are used to detect light rays and gamma radiation.

(i) Name a detector of gamma rays.   1

(ii) Complete the sentences by circling the correct words.   1

Compared to gamma rays, light rays have a $\left\{ \begin{array}{c} \text{higher} \\ \text{lower} \end{array} \right\}$ frequency

which means they have a $\left\{ \begin{array}{c} \text{higher} \\ \text{lower} \end{array} \right\}$ energy.

Total marks   **6**

MARKS | DO NOT WRITE IN THIS MARGIN

14. What affects how long it takes objects fall to the ground?

Use your knowledge of physics to answer this question.

3

[END OF MODEL QUESTION PAPER]

MARKS

ADDITIONAL SPACE FOR ROUGH WORKING AND ANSWERS

MARKS DO NOT WRITE IN THIS MARGIN

ADDITIONAL SPACE FOR ROUGH WORKING AND ANSWERS

MARKS DO NOT WRITE IN THIS MARGIN

ADDITIONAL SPACE FOR ROUGH WORKING AND ANSWERS

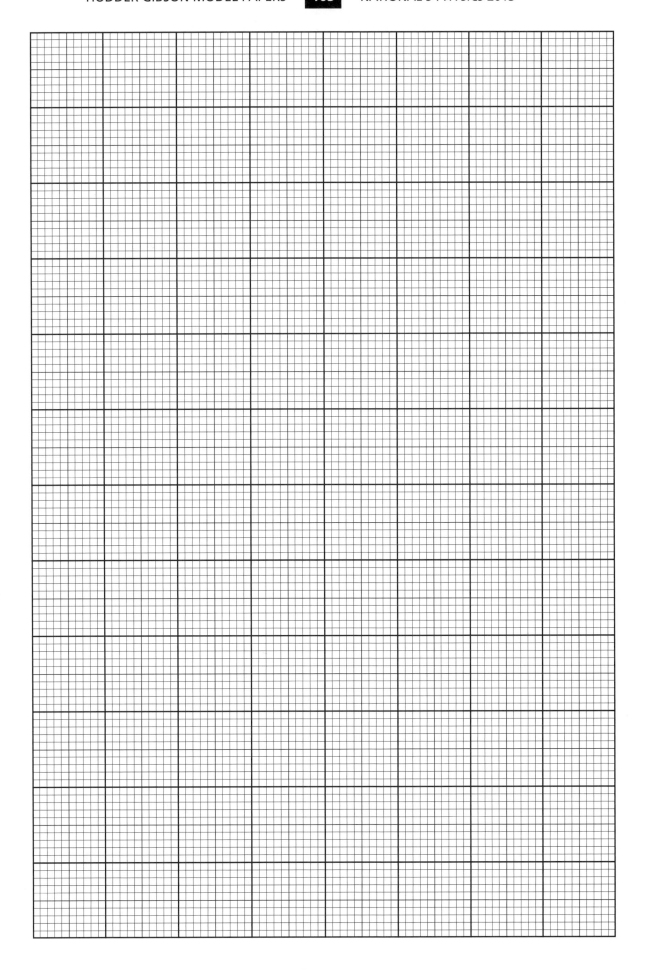

NATIONAL 5

2013 Model Paper 2

National
Qualifications
MODEL PAPER 2

Physics
Section 1—Questions

Date — Not applicable

Duration — 2 hours

Instructions for completion of Section 1 are given on Page two of the question paper.

Record your answers on the grid on Page three of your answer booklet

Do NOT write in this booklet.

Before leaving the examination room you must give your answer booklet to the Invigilator. If you do not, you may lose ALL the marks for this paper.

DATA SHEET

Speed of light in materials

Material	Speed in m s^{-1}
Air	3.0×10^8
Carbon dioxide	3.0×10^8
Diamond	1.2×10^8
Glass	2.0×10^8
Glycerol	2.1×10^8
Water	2.3×10^8

Speed of sound in materials

Material	Speed in m s^{-1}
Aluminium	5200
Air	340
Bone	4100
Carbon dioxide	270
Glycerol	1900
Muscle	1600
Steel	5200
Tissue	1500
Water	1500

Gravitational field strengths

	Gravitational field strength on the surface in N kg^{-1}
Earth	9.8
Jupiter	23
Mars	3.7
Mercury	3.7
Moon	1.6
Neptune	11
Saturn	9.0
Sun	270
Uranus	8.7
Venus	8.9

Specific heat capacity of materials

Material	Specific heat capacity in J kg^{-1} °C^{-1}
Alcohol	2350
Aluminium	902
Copper	386
Glass	500
Ice	2100
Iron	480
Lead	128
Oil	2130
Water	4180

Specific latent heat of fusion of materials

Material	Specific latent heat of fusion in J kg^{-1}
Alcohol	0.99×10^5
Aluminium	3.95×10^5
Carbon Dioxide	1.80×10^5
Copper	2.05×10^5
Iron	2.67×10^5
Lead	0.25×10^5
Water	3.34×10^5

Melting and boiling points of materials

Material	Melting point in °C	Boiling point in °C
Alcohol	−98	65
Aluminium	660	2470
Copper	1077	2567
Glycerol	18	290
Lead	328	1737
Iron	1537	2737

Specific latent heat of vaporisation of materials

Material	Specific latent heat of vaporisation in J kg^{-1}
Alcohol	11.2×10^5
Carbon Dioxide	3.77×10^5
Glycerol	8.30×10^5
Turpentine	2.90×10^5
Water	22.6×10^5

Radiation weighting factors

Type of radiation	Radiation weighting factor
alpha	20
beta	1
fast neutrons	10
gamma	1
slow neutrons	3

SECTION 1

1. A student makes the following statements about electric fields.

   I   There is a force on a charge in an electric field.

   II   When an electric field is applied to a conductor, the free electric charges in the conductor move.

   III Work is done when a charge is moved in an electric field.

   Which of these statements is/are correct?

   A   I only

   B   II only

   C   I and II only

   D   I and III only

   E   I, II and III

2. The circuit below shows a 20 V supply connected across two resistors.

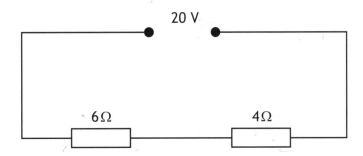

   The charge passing through the 6 Ω resistor in 3 s is

   A   2 C

   B   6 C

   C   8 C

   D   12 C

   E   20 C.

3. Three resistors are connected as shown.

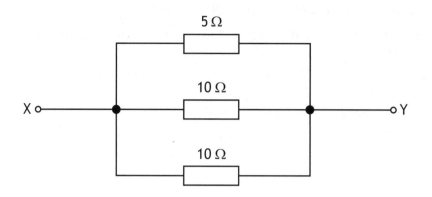

The resistance between X and Y is

A   0·06 Ω

B   0·4 Ω

C   2·5 Ω

D   13 Ω

E   25 Ω.

4. Resistors are connected in the following circuit as shown.

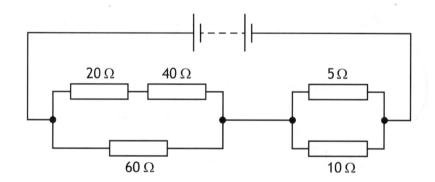

In which resistor is the current smallest?

A   5 Ω

B   10 Ω

C   20 Ω

D   40 Ω

E   60 Ω.

5. A car headlamp bulb has 24 W, 12 V printed on it.

What is the current in the bulb and the resistance of the filament when operating normally?

	Normal working current (A)	Resistance of the filament (Ω)
A	0·5	24
B	0·5	6
C	2	6
D	2	24
E	2	48

6. A temperature of 273 °C is the same as a temperature of

A   0 K

B   100 K

C   273 K

D   373 K

E   546 K.

7. The International Space Station satellite has a period of 103 minutes and an orbital height of 400km.

The GEOS–15 metrological satellite has a period of 1436 minutes and an orbital height of 36 000km.

Which of the following gives the period of the GLONASS global positioning satellite which has an orbital height of 19 000 km?

A   82 minutes

B   103 minutes

C   675 minutes

D   1436 minutes

E   1539 minutes.

8. Which of the following electromagnetic waves has a higher frequency than microwaves and a lower frequency than visible light?

    A   Gamma rays

    B   Infrared

    C   Radio

    D   Ultraviolet

    E   X-rays.

9. A ray of light passes from air into a glass block as shown.

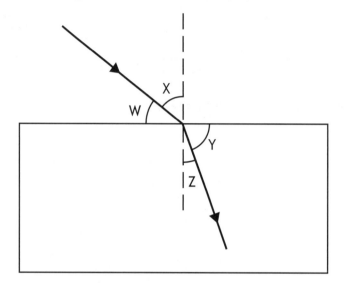

Which row in the table shows the angle of incidence and the angle of refraction?

	Angle of incidence	Angle of refraction
A	W	Z
B	W	Y
C	X	Z
D	X	Y
E	Z	X

10. A student makes the following statements.

I   The nucleus of an atom contains protons and electrons.

II   Gamma radiation produces the greatest ionisation density.

III Beta particles are fast moving electrons.

Which of these statements is/are correct?

A   I only

B   I and III only

C   II only

D   II and III only

E   III only

11. A radioactive source emits alpha, beta and gamma radiation. A detector, connected to a counter, is placed 10 mm in front of the source. The counter records 400 counts per minute.

A sheet of paper is placed between the source and the detector. The counter records 300 counts per minute.

The radiation now detected is

A   alpha only

B   alpha and beta only

C   beta only

D   beta and gamma only

E   gamma only.

12. A radioactive tracer is a liquid which is injected into a patient to study the flow of blood.

The radioactive tracer is carried around the body by the patient's blood.

A detector is placed above the patient to monitor the flow of blood carrying the tracer.

The tracer should have a

A   short half-life and emit α particles

B   long half-life and emit β particles

C   long half-life and emit γ rays

D   long half-life and emit α particles

E   short half-life and emit γ rays.

13. For a particular radioactive source, 1800 atoms decay in a time of 3 minutes. The **activity** of this source is

A   10 Bq

B   600 Bq

C   800 Bq

D   5400 Bq

E   324 000 Bq.

14. Human tissue can be damaged by exposure to radiation.

On which of the following factors does the risk of biological harm depend?

I   The absorbed dose.

II   The type of radiation.

III   The body organs or tissue exposed.

A   I only

B   I and II only

C   II only

D   II and III only

E   I, II and III

15. At an airport an aircraft moves from the terminal building to the end of the runway.

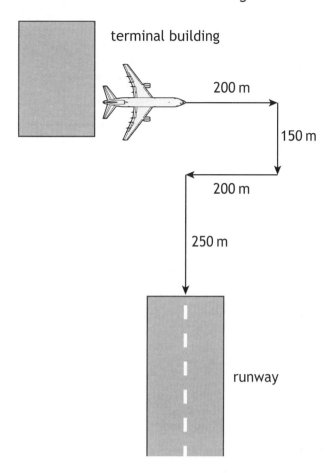

Which row shows the total distance travelled and the size of the displacement of the aircraft?

	Total distance travelled (m)	Size of displacement (m)
A	400	800
B	450	200
C	450	400
D	800	400
E	800	800

16. A block is pulled across a horizontal surface as shown.

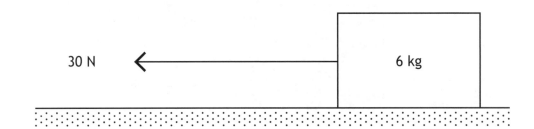

The mass of the block is 6 kg.

The block is travelling at constant speed.

The force of friction acting on the block is

A   0 N

B   5 N

C   24 N

D   30 N

E   36 N.

17. A space probe with a mass of 760 kg landed on the surface of a planet in our solar system. The weight of the probe at the surface of the planet in our solar system was 6764 N.

The planet was

A   Jupiter

B   Mars

C   Neptune

D   Saturn

E   Venus.

18. Near the Earth's surface, a mass of 5 kg is falling with a constant velocity.

The air resistance and the unbalanced force acting on the mass are:

	air resistance	unbalanced force
A	49 N upwards	49 N downwards
B	9·8 N upwards	9·8 N downwards
C	9·8 N downwards	58·8 N downwards
D	9·8 N upwards	0 N
E	49 N upwards	0 N

19. Two identical balls X and Y are projected horizontally from the edge of a cliff. The path taken by each ball is shown.

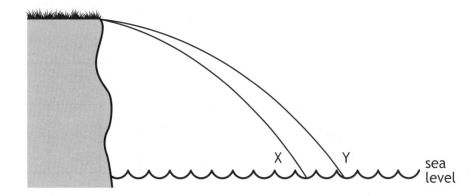

A student makes the following statements about the motion of the two balls.

I   They take the same time to reach sea level.

II   They have the same vertical acceleration.

III They have the same horizontal velocity.

Which of these statements is/are correct?

A   I only

B   II only

C   I and II only

D   I and III only

E   II and III only

20.

The line spectrum for an element is shown below.

Line spectrum of element

The line spectra of different stars are shown below.
Identify which star this element is present in.

A

B

C

D

E

[END OF SECTION 1. NOW ATTEMPT THE QUESTIONS IN SECTION 2
OF YOUR QUESTION AND ANSWER BOOKLET]

National Qualifications
MODEL PAPER 2

Physics
Relationships Sheet

Date — Not applicable

$$E_p = mgh$$

$$E_k = \tfrac{1}{2}mv^2$$

$$Q = It$$

$$V = IR$$

$$R_T = R_1 + R_2 + \ldots$$

$$\frac{1}{R_T} = \frac{1}{R_1} + \frac{1}{R_2} + \ldots$$

$$V_2 = \left(\frac{R_2}{R_1 + R_2}\right)V_s$$

$$\frac{V_1}{V_2} = \frac{R_1}{R_2}$$

$$P = \frac{E}{t}$$

$$P = IV$$

$$P = I^2 R$$

$$P = \frac{V^2}{R}$$

$$E_h = cm\Delta T$$

$$p = \frac{F}{A}$$

$$\frac{pV}{T} = \text{constant}$$

$$p_1 V_1 = p_2 V_2$$

$$\frac{p_1}{T_1} = \frac{p_2}{T_2}$$

$$\frac{V_1}{T_1} = \frac{V_2}{T_2}$$

$$d = vt$$

$$v = f\lambda$$

$$T = \frac{1}{f}$$

$$A = \frac{N}{t}$$

$$D = \frac{E}{m}$$

$$H = Dw_R$$

$$\dot{H} = \frac{H}{t}$$

$$s = vt$$

$$d = \bar{v}t$$

$$s = \bar{v}t$$

$$a = \frac{v - u}{t}$$

$$W = mg$$

$$F = ma$$

$$E_w = Fd$$

$$E_h = ml$$

[END OF SPECIMEN RELATIONSHIPS SHEET]

Page two

National Qualifications MODEL PAPER 2

Physics Section 1— Answer Grid and Section 2

Duration — 2 hours

Total marks — 110

SECTION 1 — 20 marks

Attempt ALL questions in this section.

Instructions for completion of Section 1 are given on Page two.

SECTION 2 — 90 marks

Attempt ALL questions in this section.

Read all questions carefully before answering.

Use **blue** or **black** ink. Do NOT use gel pens.

Write your answers in the spaces provided. Additional space for answers and rough work is provided at the end of this booklet. If you use this space, write clearly the number of the question you are answering. Any rough work must be written in this booklet. You should score through your rough work when you have written your fair copy.

Before leaving the examination room you must give this booklet to the Invigilator. If you do not, you may lose all the marks for this paper.

SECTION 1 — 20 marks

The questions for Section 1 are contained in the booklet Physics Section 1 — Questions. Read these and record your answers on the grid on Page three opposite.

1. The answer to each question is **either** A, B, C, D or E. Decide what your answer is, then fill in the appropriate bubble (see sample question below).

2. There is **only one correct** answer to each question.

3. Any rough working should be done on the rough working sheet.

Sample Question

The energy unit measured by the electricity meter in your home is the:

    A   ampere

    B   kilowatt-hour

    C   watt

    D   coulomb

    E   volt.

The correct answer is **B**—kilowatt-hour. The answer **B** bubble has been clearly filled in (see below).

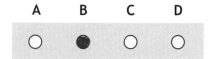

Changing an answer

If you decide to change your answer, cancel your first answer by putting a cross through it (see below) and fill in the answer you want. The answer below has been changed to **D**.

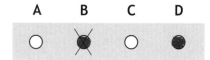

If you then decide to change back to an answer you have already scored out, put a tick (✓) to the **right** of the answer you want, as shown below:

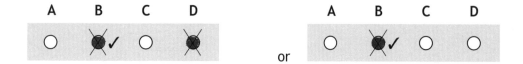

or

SECTION 1 — Answer Grid

	A	B	C	D	E
1	○	○	○	○	○
2	○	○	○	○	○
3	○	○	○	○	○
4	○	○	○	○	○
5	○	○	○	○	○
6	○	○	○	○	○
7	○	○	○	○	○
8	○	○	○	○	○
9	○	○	○	○	○
10	○	○	○	○	○
11	○	○	○	○	○
12	○	○	○	○	○
13	○	○	○	○	○
14	○	○	○	○	○
15	○	○	○	○	○
16	○	○	○	○	○
17	○	○	○	○	○
18	○	○	○	○	○
19	○	○	○	○	○
20	○	○	○	○	○

[BLANK PAGE]

SECTION 2 — 90 marks

Attempt ALL questions

MARKS | DO NOT WRITE IN THIS MARGIN

1. An early method of crash testing involved a car rolling down a slope and colliding with a wall.

   In one test, a car of mass 750 kg starts at the top of a 7·2 m high slope.

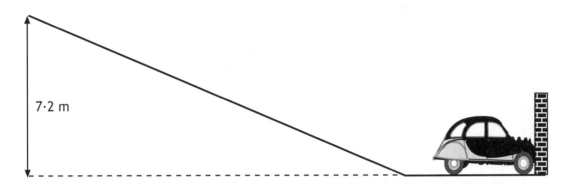

7·2 m

   (a) Calculate the gravitational potential energy of the car at the top of the slope.

   Space for working and answer.

   3

   (b) (i) State the value of the kinetic energy of the car at the bottom of the slope, assuming no energy losses.

   1

   (ii) Calculate the speed of the car at the bottom of the slope, before hitting the wall.

   Space for working and answer.

   3

   Total marks   7

MARKS | DO NOT WRITE IN THIS MARGIN

2. An office has an automatic window blind that closes when the light level outside gets too high.

The electronic circuit that operates the motor to close the blind is shown.

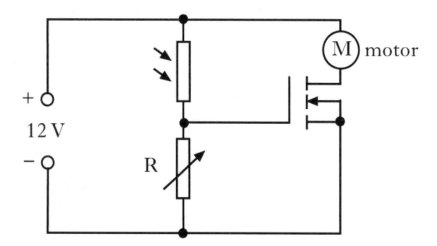

(a) The MOSFET transistor switches on when the voltage across variable resistor R reaches 2·4 V.

Explain how this circuit works to close the blind.

2

(b) The graph shows how the resistance of the LDR varies with light level.

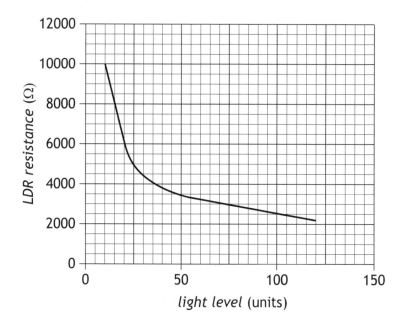

MARKS

2. (continued)

    (i) What is the resistance of the LDR when the light level is 70 units?   **1**

    (ii) R has a value of 600 Ω.

        Calculate the voltage across R when the light level is 70 units.   **3**

        Space for working and answer.

    (iii) State whether or not the blinds will close when the light level is 70 units.

        You must justify your answer.   **2**

Total marks   **9**

MARKS | DO NOT WRITE IN THIS MARGIN

3. (a) State the definition of pressure.

1

(b) To remain safe when diving, deep sea divers must know the pressure exerted on them by the sea at different depths.

The pressure exerted on deep sea divers when diving beneath the sea is calculated using the relationship:

$$p = \rho gh$$

where:

p is the pressure exerted by the sea in Pa

ρ is the density of the sea water in kg m^{-3}

g is the acceleration due to gravity in m s^{-2}

h is the submerged depth of the diver in metres.

During one dive, a diver reaches a depth of 24 m. The density of the water is 1025 kg m^{-3}.

Calculate the pressure exerted by the sea water on the diver at this depth.   **3**

Space for working and answer.

Total marks   4

MARKS | DO NOT WRITE IN THIS MARGIN

4. Car designers are constantly trying to reduce the environmental impact of cars. One way to do this is to make them more fuel-efficient, as the less fuel cars need, the less dangerous gases they emit into the atmosphere.

Use your knowledge of physics to comment on how car manufacturers might produce cars which are more fuel efficient.

3

5. A student carries out an experiment to investigate the relationship between the pressure and temperature of a fixed mass of gas. The apparatus used is shown.

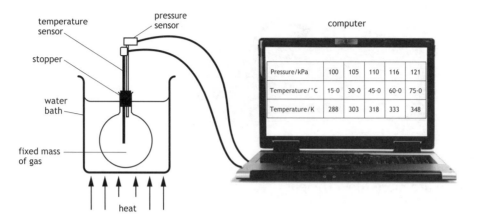

The pressure and temperature of the gas are recorded using sensors connected to a computer. The gas is heated slowly in the water bath and a series of readings is taken.

The volume of the gas remains constant during the experiment.

The results are shown.

Pressure/kPa	100	105	110	116	121
Temperature/°C	15·0	30·0	45·0	60·0	75·0
Temperature/K	288	303	318	333	348

MARKS

5. (continued)

(a) Using **all** the relevant data, establish the relationship between the pressure and the temperature of the gas.

2

(b) Use the kinetic model to explain the change in pressure as the temperature of the gas increases.

3

(c) Explain why the level of water in the water bath should be above the bottom of the stopper.

1

Total marks 6

MARKS | DO NOT WRITE IN THIS MARGIN

6. A satellite sends microwaves to a ground station on Earth.

(a) The microwaves have a wavelength of 60mm.

   (i) Calculate the frequency of the waves.   **3**
      Space for working and answer.

   (ii) Calculate the period of the waves.   **3**
      Space for working and answer.

(b) The satellite sends radio waves along with the microwaves to the ground station. Will the radio waves be received by the ground station **before**, **after** or **at the same time** as the microwaves?

   Explain your answer.   **2**

MARKS | DO NOT WRITE IN THIS MARGIN

6. **(continued)**

(c) A music concert is being broadcast live on radio.

Drivers in two cars, A and B, are listening to the performance on the radio.

radio
transmitter

The performance is being broadcast on two different wavebands, from the same transmitter.

The radio in car A is tuned to a radio signal of frequency 1152 kHz.

The radio in car B is tuned to a radio signal of frequency 102·5 MHz.

Both cars drive into a valley surrounded by hills.

The radio in car B loses the signal from the broadcast.

Explain why this signal is lost. 2

Total marks 10

MARKS | DO NOT WRITE IN THIS MARGIN

7. Gold-198 is a radioactive source that is used to trace factory waste which may cause river pollution.

A small quantity of the radioactive gold is added into the waste as it enters the river. Scanning the river using radiation detectors allows scientists to trace where the waste has travelled.

Gold-198 has a half-life of 2·7 days.

(a) What is meant by the term "half-life"?   1

(b) A sample of Gold-198 has an activity of 64kBq when first obtained by the scientists.

Calculate the activity after 13·5 days.   3

Space for working and answer.

(c) Describe two precautions taken by the scientists to reduce the equivalent dose they receive while using radioactive sources.   2

Total marks   6

MARKS | DO NOT WRITE IN THIS MARGIN

8. Many countries use nuclear reactors to produce energy.

Nuclear fission occurs inside the reactor.

A diagram of the core of a nuclear reactor is shown.

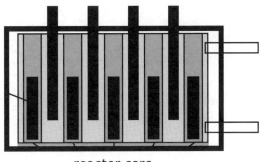

reactor core

(a) Describe what happens when a nuclear fission reaction occurs.   2

(b) One nuclear fission reaction produces $2 \cdot 9 \times 10^{-11}$ J of energy.

The power output of the reactor is $1 \cdot 4$ GW.

How many fission reactions are produced in one hour?   4
Space for working and answer.

(c) State **one disadvantage** of using nuclear power for the generation of electricity.   1

Total marks   7

9. An aircraft is flying horizontally at a constant speed.

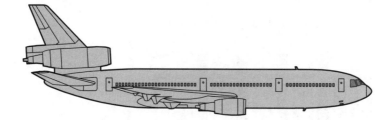

(a) The aircraft and passengers have a total mass of 50 000 kg.

Calculate the total weight.

Space for working and answer.

3

(b) State the magnitude of the upward force acting on the aircraft.

1

MARKS | DO NOT WRITE IN THIS MARGIN

9.   (continued)

(c)   During the flight, the aircraft's engines produce a force of 4.4×10^4 N due North. The aircraft encounters a crosswind, blowing from west to east, which exerts a force of 3.2×10^4 N.

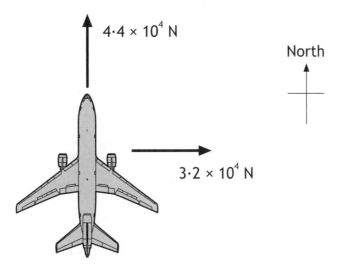

Calculate the resultant force on the aircraft.   4

Space for working and answer.

MARKS | DO NOT WRITE IN THIS MARGIN

9. **(continued)**

(d) During a particular flight, a pilot receives an absorbed dose of 15 µGy from gamma rays.

Calculate the equivalent dose received due to this type of radiation.   **4**

Space for working and answer.

(e) Gamma radiation is an example of radiation which causes ionisation.

Explain what is meant by the term ionisation.   **1**

Total marks 12

MARKS

10. Athletes in a race are recorded by a TV camera which runs on rails beside the track.

camera

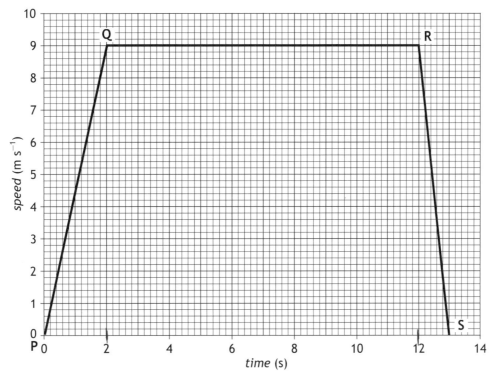

The graph shows the speed of the camera during the race.

(a) Calculate the acceleration of the camera between **P** and **Q**.

Space for working and answer.

3

MARKS

DO NOT WRITE IN THIS MARGIN

10. **(continued)**

(b) The mass of the camera is 15 kg.

Calculate the unbalanced force needed to produce the acceleration between **P** and **Q**.   3

Space for working and answer.

(c) How far does the camera travel in the 13 s?   3

Space for working and answer.

Total marks   9

MARKS | DO NOT WRITE IN THIS MARGIN

11. One type of exercise machine is shown below.

(a) A person using this machine pedals against friction forces applied to the wheel by the brake.

A friction force of 300 N is applied at the edge of the wheel, which has a circumference of 1.5m.

How much work is done by friction in 500 turns of the wheel?   **4**

Space for working and answer.

MARKS

11. (continued)

(b) The wheel is a solid aluminium disc of mass 12.0kg.

(i) All the work done by friction is converted to heat in the disc.

Calculate the temperature rise after 500 turns.   4

Space for working and answer.

(ii) Explain why the actual temperature rise of the disc is less than calculated in (b)(i)   1

Total marks   9

MARKS | DO NOT WRITE IN THIS MARGIN

12. When a spacecraft is launched into space it accelerates to reach speeds of up to 8 km s^{-1} to achieve orbit.

At launch, most of its mass consists of the fuel required to provide upthrust for this acceleration.

During the launch, the acceleration of the spacecraft is not constant.

Use your knowledge of physics to comment on why the acceleration is not constant.

3

MARKS | DO NOT WRITE IN THIS MARGIN

13.   Read the passage below and answer the questions that follow.

Neutron stars

When stars reach the end of their life, they can become Neutron stars. A neutron star has a mass ranging from 1·4-3·2 times that of our Sun. In a neutron star, this huge mass is contained within a diameter of approximately 12 km, and means that a neutron star is extremely dense.

One cupful of this mass would have the same weight as Mount Everest on Earth!

The extreme density of the neutron star also means that it has very large gravitational and magnetic field strengths.

Stars emit huge amounts of energy. This energy is the result of nuclear fusion happening at the centre of the star. Nuclear fusion of the isotopes of hydrogen produces helium and also the energy which sustains the star's massive shape.

Neutron stars are thought to be formed when large stars collapse.

This happens when the fusion process stops and there is no longer enough energy to sustain the star. The star explodes.

This explosion is known as a supernova. The outer gases of the star expand rapidly to produce an extremely bright object in the sky, which can be seen by astronomers on Earth.

The gravitational field causes the centre of the star to collapse. Its volume reduces dramatically. During the collapsing process, electrons and protons combine to form neutrons. This is the reason for the name 'Neutron' stars.

Neutron stars sometimes appear in binary systems, where they are in mutual orbit around another object. X-ray telescopes on satellites have been used by astronomers to obtain data from such binary systems. This data has confirmed the mass of the Neutron star to be 1·4-3·2 times that of the Sun's mass.

Neutron stars rotate rapidly when newly formed, and gradually slow over a long period of time. A neutron star, known as PSR J1748-2446ad, rotates 716 times per second.

Some neutron stars emit radio waves or X-rays. These emissions only occur at the magnetic poles of the neutron star. When observed by astronomers, these emissions appear as 'pulses' of radio waves or X-rays. The pulses appear at the same rate as the rotation of the neutron star. Such neutron stars are known as 'Pulsars'.

MARKS

DO NOT
WRITE IN
THIS
MARGIN

13. **(continued)**

(a) What event is thought to lead to the formation of neutron stars?   1

(b) Why does the neutron star consist mainly of neutrons?   1

(c) Calculate the period of rotation of the neutron star known as PSR J1748-2446ad.   3

Space for working and answer.

Total marks   5

[END OF MODEL QUESTION PAPER]

MARKS

ADDITIONAL SPACE FOR ROUGH WORKING AND ANSWERS

MARKS

ADDITIONAL SPACE FOR ROUGH WORKING AND ANSWERS

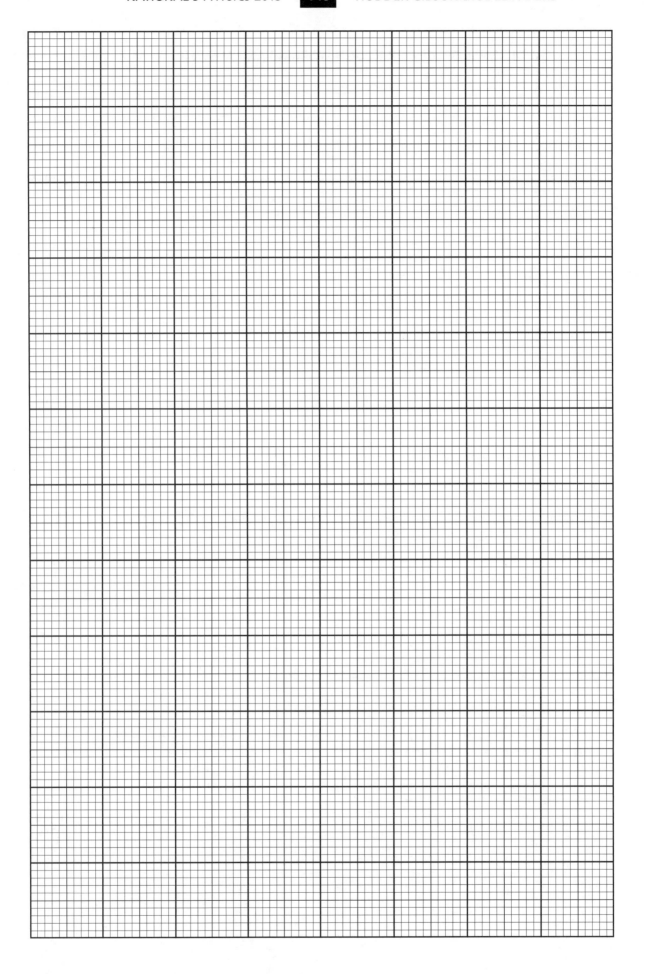

NATIONAL 5

2013 Model Paper 3

National Qualifications
MODEL PAPER 3

Physics
Section 1—Questions

Date — Not applicable

Duration — 2 hours

Instructions for completion of Section 1 are given on Page two of the question paper.

Record your answers on the grid on Page three of your answer booklet

Do NOT write in this booklet.

Before leaving the examination room you must give your answer booklet to the Invigilator. If you do not, you may lose ALL the marks for this paper.

DATA SHEET

Speed of light in materials

Material	Speed in m s^{-1}
Air	$3 \cdot 0 \times 10^8$
Carbon dioxide	$3 \cdot 0 \times 10^8$
Diamond	$1 \cdot 2 \times 10^8$
Glass	$2 \cdot 0 \times 10^8$
Glycerol	$2 \cdot 1 \times 10^8$
Water	$2 \cdot 3 \times 10^8$

Speed of sound in materials

Material	Speed in m s^{-1}
Aluminium	5200
Air	340
Bone	4100
Carbon dioxide	270
Glycerol	1900
Muscle	1600
Steel	5200
Tissue	1500
Water	1500

Gravitational field strengths

	Gravitational field strength on the surface in N kg^{-1}
Earth	$9 \cdot 8$
Jupiter	23
Mars	$3 \cdot 7$
Mercury	$3 \cdot 7$
Moon	$1 \cdot 6$
Neptune	11
Saturn	$9 \cdot 0$
Sun	270
Uranus	$8 \cdot 7$
Venus	$8 \cdot 9$

Specific heat capacity of materials

Material	Specific heat capacity in J kg^{-1} °C^{-1}
Alcohol	2350
Aluminium	902
Copper	386
Glass	500
Ice	2100
Iron	480
Lead	128
Oil	2130
Water	4180

Specific latent heat of fusion of materials

Material	Specific latent heat of fusion in J kg^{-1}
Alcohol	$0 \cdot 99 \times 10^5$
Aluminium	$3 \cdot 95 \times 10^5$
Carbon Dioxide	$1 \cdot 80 \times 10^5$
Copper	$2 \cdot 05 \times 10^5$
Iron	$2 \cdot 67 \times 10^5$
Lead	$0 \cdot 25 \times 10^5$
Water	$3 \cdot 34 \times 10^5$

Melting and boiling points of materials

Material	Melting point in °C	Boiling point in °C
Alcohol	−98	65
Aluminium	660	2470
Copper	1077	2567
Glycerol	18	290
Lead	328	1737
Iron	1537	2737

Specific latent heat of vaporisation of materials

Material	Specific latent heat of vaporisation in J kg^{-1}
Alcohol	$11 \cdot 2 \times 10^5$
Carbon Dioxide	$3 \cdot 77 \times 10^5$
Glycerol	$8 \cdot 30 \times 10^5$
Turpentine	$2 \cdot 90 \times 10^5$
Water	$22 \cdot 6 \times 10^5$

Radiation weighting factors

Type of radiation	Radiation weighting factor
alpha	20
beta	1
fast neutrons	10
gamma	1
slow neutrons	3

SECTION 1

1. The unit of current is the ampere.

    One ampere can also be expressed as

    A   one volt per joule

    B   one coulomb per second

    C   one ohm per volt

    D   one joule per second

    E   one joule per coulomb.

2. The potential difference across a lamp is 8 V when the current is 2 A.

    The charge which passes through the lamp in three minutes is

    A   6 C

    B   16 C

    C   18 C

    D   360 C

    E   1080 C.

3. Which diagram correctly shows the direction which atomic particles electrons, protons and neutrons are deflected by an electric field?

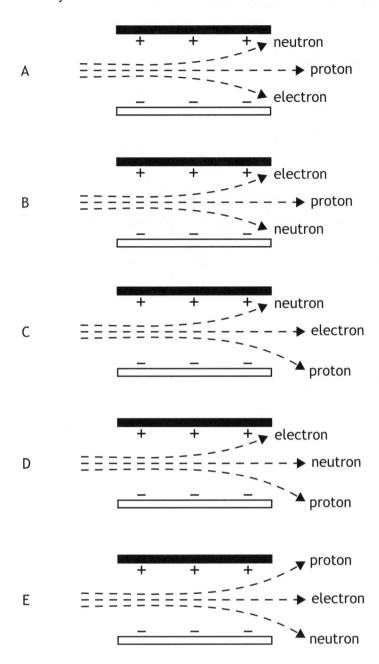

4. The graph below shows how the current is related to the applied potential difference for two separate resistors P and Q.

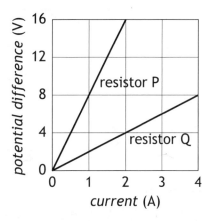

The values of the two resistors shown, in ohms, are

	Resistance of P (Ω)	Resistance of Q (Ω)
A	0·125	0·5
B	2	4
C	8	2
D	8	16
E	16	8

5. A circuit is set up as shown.

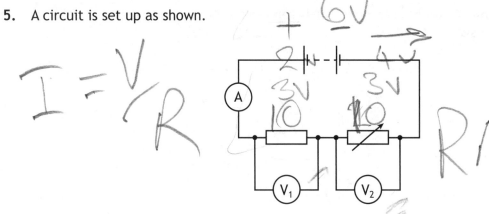

The resistance of the variable resistor is increased.

Which row in the table shows the effect on the readings on the ammeter and voltmeters?

	Reading on ammeter	Reading on voltmeter V₁	Reading on voltmeter V₂
A	decreases	decreases	decreases
B	increases	unchanged	increases
C	decreases	increases	decreases
D	increases	unchanged	decreases
E	decreases	decreases	increases

6.   A circuit is set up as shown.

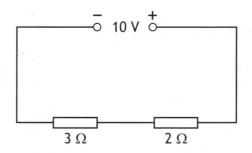

The potential difference across the 2 Ω resistor is

A   4 V

B   5 V

C   6 V

D   10 V

E   20 V.

7.   Which of the following describes the frequency of a water wave?

A   The distance between the crest of a wave and the crest of the next wave

B   The time taken for one complete wave to pass any point

C   The number of waves passing any point in one second

D   The distance travelled by a crest in one second

E   The time taken for the source to make one complete vibration

8. Water waves are produced in two identical ripple tanks. The waves reach a barrier and are diffracted.

   Which pair of ripple tanks shows correct diffraction?

A
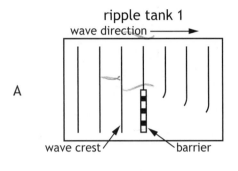
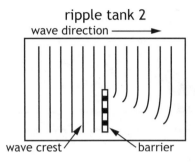

B
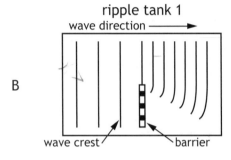
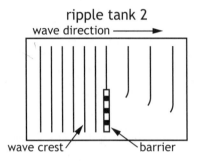

C
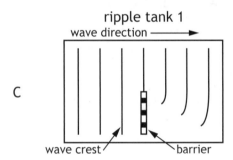
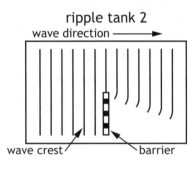

D
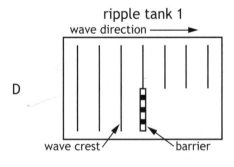
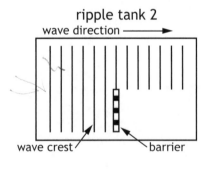

E
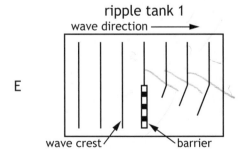
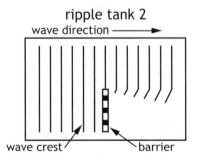

9. Which of the following electromagnetic waves has a higher frequency than visible light and a lower frequency than X-rays?

    A   Gamma rays

    B   Infrared

    C   Microwaves

    D   Radio

    E   Ultraviolet

10. Which of the following describes the term ionisation?

    A   An atom losing an orbiting electron

    B   An atom losing a proton

    C   A nucleus emitting an alpha particle

    D   A nucleus emitting a neutron

    E   A nucleus emitting a gamma ray

11. Compared with a proton, an alpha particle has

    A   twice the mass and twice the charge

    B   twice the mass and the same charge

    C   four times the mass and twice the charge

    D   four times the mass and the same charge

    E   twice the mass and four times the charge.

12. One gray is equal to

    A   one becquerel per kilogram

    B   one sievert per second

    C   one joule per second

    D   one sievert per kilogram

    E   one joule per kilogram.

13. A sample of a radioactive material has a mass of 30 g. There are 36 000 nuclear decays every minute in this sample.

The activity of the sample is

A   600 Bq

B   1800 Bq

C   36 000 Bq

D   1 080 000 Bq

E   2 160 000 Bq.

14. Which of the following statements about nuclear fusion is correct?

A   Energy is released when a nucleus with a large mass number splits into two nuclei of smaller mass number.

B   Energy is absorbed when a nucleus with a large mass number splits into two nuclei of smaller mass number.

C   Energy is absorbed when two nuclei combine to form a nucleus of larger mass number.

D   Energy is released when two nuclei combine to form a nucleus of larger mass number.

E   Energy is absorbed when a nucleus with a large mass number combines with a nucleus of small mass number to produce a nucleus of larger mass number.

15. Which row contains two scalar quantities and one vector quantity?

A   Distance, weight, velocity

B   Speed, mass, displacement

C   Distance, weight, force

D   Speed, weight, acceleration

E   Velocity, force, mass

16. A student follows the route shown in the diagram and arrives back at the starting point.

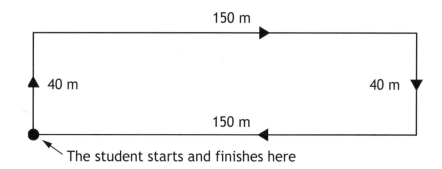

150 m

40 m   40 m

150 m

The student starts and finishes here

Which row in the table shows the total distance walked and the magnitude of the final displacement?

	Total distance (m)	Final displacement (m)
A	0	80
B	0	380
C	190	0
D	380	0
E	380	380

17. The graph shows how the velocity of an object varies with time.

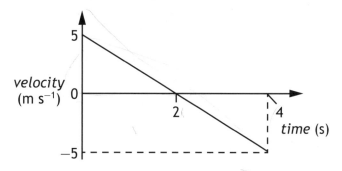

velocity (m s⁻¹)

5

0

2   4

−5

time (s)

Which row in the table shows the displacement after 4 s and the acceleration of the object during the first 4 s?

	Displacement (m)	Acceleration (m s⁻²)
A	10	−10
B	10	2·5
C	0	2·5
D	0	−10
E	0	−2·5

18. A ball was dropped into a deep well. The graph shows the speed of the ball from the instant of its release in air until it has fallen several metres through the water to the bottom of the well.

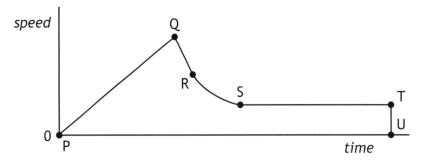

Which part of the graph indicates when the vertical forces acting on the ball were balanced?

A   PQ

B   QR

C   RS

D   ST

E   TU

19. A package is released from a helicopter flying horizontally at a constant speed of 39 m s^{-1}.

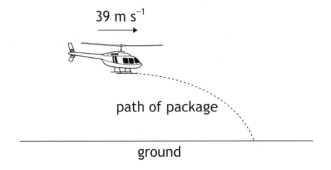

The package takes 5·0 s to reach the ground.

The effects of air resistance can be ignored.

Which row in the table shows the horizontal speed and vertical speed of the package just before it hits the ground?

	Horizontal speed (m s^{-1})	Vertical speed (m s^{-1})
A	0	39
B	39	39
C	39	49
D	49	39
E	49	49

20. The diameter of the disk of the Milky Way galaxy spans a distance of about 100,000 light years.

The diameter of the Milky Way galaxy, expressed in metres is

A $9{\cdot}5{\times}10^5$

B $3{\cdot}0{\times}10^{13}$

C $1{\cdot}1{\times}10^{16}$

D $2{\cdot}6{\times}10^{17}$

E $9{\cdot}5{\times}10^{20}$.

**[END OF SECTION 1. NOW ATTEMPT THE QUESTIONS IN SECTION 2
OF YOUR QUESTION AND ANSWER BOOKLET]**

[BLANK PAGE]

National Qualifications
MODEL PAPER 3

Physics
Relationships Sheet

Date — Not applicable

$$E_p = mgh$$

$$E_k = \frac{1}{2}mv^2$$

$$Q = It$$

$$V = IR$$

$$R_T = R_1 + R_2 + \ldots$$

$$\frac{1}{R_T} = \frac{1}{R_1} + \frac{1}{R_2} + \ldots$$

$$V_2 = \left(\frac{R_2}{R_1 + R_2}\right)V_s$$

$$\frac{V_1}{V_2} = \frac{R_1}{R_2}$$

$$P = \frac{E}{t}$$

$$P = IV$$

$$P = I^2 R$$

$$P = \frac{V^2}{R}$$

$$E_h = cm\Delta T$$

$$p = \frac{F}{A}$$

$$\frac{pV}{T} = \text{constant}$$

$$p_1 V_1 = p_2 V_2$$

$$\frac{p_1}{T_1} = \frac{p_2}{T_2}$$

$$\frac{V_1}{T_1} = \frac{V_2}{T_2}$$

$$d = vt$$

$$v = f\lambda$$

$$T = \frac{1}{f}$$

$$A = \frac{N}{t}$$

$$D = \frac{E}{m}$$

$$H = Dw_R$$

$$\dot{H} = \frac{H}{t}$$

$$s = vt$$

$$d = \bar{v}t$$

$$s = \bar{v}t$$

$$a = \frac{v - u}{t}$$

$$W = mg$$

$$F = ma$$

$$E_w = Fd$$

$$E_h = ml$$

[END OF SPECIMEN RELATIONSHIPS SHEET]

Page two

National Qualifications MODEL PAPER 3

Physics Section 1— Answer Grid and Section 2

Duration — 2 hours

Total marks — 110

SECTION 1 — 20 marks

Attempt ALL questions in this section.

Instructions for completion of Section 1 are given on Page two.

SECTION 2 — 90 marks

Attempt ALL questions in this section.

Read all questions carefully before answering.

Use **blue** or **black** ink. Do NOT use gel pens.

Write your answers in the spaces provided. Additional space for answers and rough work is provided at the end of this booklet. If you use this space, write clearly the number of the question you are answering. Any rough work must be written in this booklet. You should score through your rough work when you have written your fair copy.

Before leaving the examination room you must give this booklet to the Invigilator. If you do not, you may lose all the marks for this paper.

SECTION 1 — 20 marks

The questions for Section 1 are contained in the booklet Physics Section 1 — Questions.
Read these and record your answers on the grid on Page three opposite.

1. The answer to each question is **either** A, B, C, D or E. Decide what your answer is, then fill in the appropriate bubble (see sample question below).

2. There is **only one correct** answer to each question.

3. Any rough working should be done on the rough working sheet.

Sample Question

The energy unit measured by the electricity meter in your home is the:

    A   ampere

    B   kilowatt-hour

    C   watt

    D   coulomb

    E   volt.

The correct answer is **B**—kilowatt-hour. The answer **B** bubble has been clearly filled in (see below).

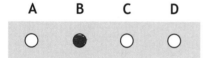

Changing an answer

If you decide to change your answer, cancel your first answer by putting a cross through it (see below) and fill in the answer you want. The answer below has been changed to **D**.

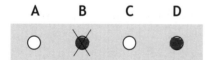

If you then decide to change back to an answer you have already scored out, put a tick (✓) to the **right** of the answer you want, as shown below:

    or

SECTION 1 — Answer Grid

	A	B	C	D	E
1	○	○	○	○	○
2	○	○	○	○	○
3	○	○	○	○	○
4	○	○	○	○	○
5	○	○	○	○	○
6	○	○	○	○	○
7	○	○	○	○	○
8	○	○	○	○	○
9	○	○	○	○	○
10	○	○	○	○	○
11	○	○	○	○	○
12	○	○	○	○	○
13	○	○	○	○	○
14	○	○	○	○	○
15	○	○	○	○	○
16	○	○	○	○	○
17	○	○	○	○	○
18	○	○	○	○	○
19	○	○	○	○	○
20	○	○	○	○	○

[BLANK PAGE]

SECTION 2 — 90 marks

Attempt ALL questions

MARKS | DO NOT WRITE IN THIS MARGIN

1. A student reproduces Galileo's famous experiment by dropping a solid copper ball of mass 0·50 kg from a balcony on the Leaning Tower of Pisa.

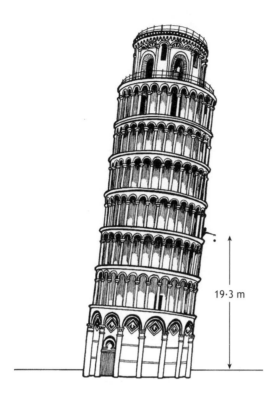

19·3 m

(a) The ball is released from a height of 19·3 m.

Calculate the gravitational potential energy lost by the ball.   **3**

Space for working and answer.

MARKS | DO NOT WRITE IN THIS MARGIN

1. (continued)

(b) Assuming that all of this gravitational potential energy is converted into heat energy **in the ball**, calculate the increase in the temperature of the ball on impact with the ground.   3

Space for working and answer.

(c) Is the actual temperature change of the ball greater than, the same as or less than the value calculated in part (a)(ii)?

You **must** explain your answer.   2

Total marks   8

MARKS | DO NOT WRITE IN THIS MARGIN

2. Some resistors are labelled with a power rating as well as their resistance value. This is the maximum power at which they can operate without overheating.

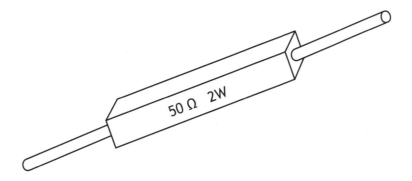

(a) A resistor is labelled 50 Ω, 2W.

Calculate the maximum operating current for this resistor.   3

Space for working and answer.

(b) Two resistors, each rated at 2W, are connected in parallel to a 9 V d.c. supply.

They have resistances of 60 Ω and 30 Ω.

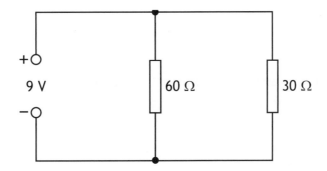

MARKS | DO NOT WRITE IN THIS MARGIN

2. (continued)

(i) Calculate the total resistance of the circuit.   3
Space for working and answer.

(ii) Calculate the power produced in each resistor.   4
Space for working and answer.

(iii) State which, if any, of the resistors will overheat.   1

(c) The 9 V **d.c.** supply is replaced by a 9 V **a.c.** supply.

What effect, if any, would this have on your answers to part (b) (ii)?   1

Total marks 12

3. A mass of copper heated with a Bunsen is immersed in a beaker of cold water.

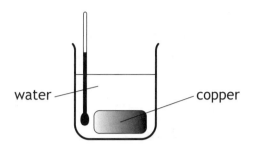

water —   — copper

Use your knowledge of physics to comment on what the final temperature of the copper and water would depend on.

Make reference to any relevant equation(s) in your answer.

3

MARKS | DO NOT WRITE IN THIS MARGIN

4. A garden spray consists of a tank, a pump and a spray nozzle.

The tank is partially filled with water.

The pump is then used to increase the pressure of the air above the water.

The volume of the compressed air in the tank is $1 \cdot 60 \times 10^{-3}$ m³.

The surface area of the water is $3 \cdot 00 \times 10^{-2}$ m².

The pressure of the air in the tank is $4 \cdot 60 \times 10^{5}$ Pa.

(a) Calculate the force on the surface of the water.   3

Space for working and answer.

MARKS | DO NOT WRITE IN THIS MARGIN

4. (continued)

(b) The spray nozzle is operated and water is pushed out until the pressure of the air in the tank is $1{\cdot}00 \times 10^5$ Pa.

Calculate the new volume of compressed air in the tank. **3**

Space for working and answer.

(c) Calculate the volume of water expelled. **1**

Space for working and answer.

Total marks 7

DO NOT
WRITE IN
THIS
MARGIN

5. A ripple tank is set up to investigate the properties of water waves.

A wave generator is used to produce the waves in the tank.

(a) When the wave generator is vibrating at 5 Hz, it is found that there are 8 complete waves between the wave generator and the opposite side of the tank, as shown in figure 1.

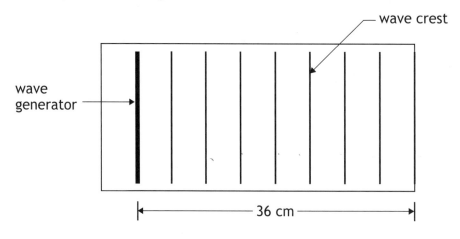

figure 1

Calculate the speed of the water waves.   4

Space for working and answer.

MARKS

5. (continued)

(b) A barrier with a wide gap in it is placed across the middle of the tank as shown in figure 2.

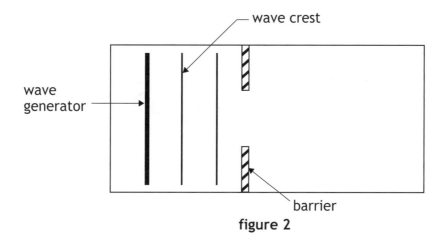

figure 2

Complete figure 2 showing the wave pattern on the right hand side of the barrier.

2

(c) Optical fibres are used to carry internet data using infra-red radiation.

The diagram shows a view of an infra-red ray entering the end of a fibre.

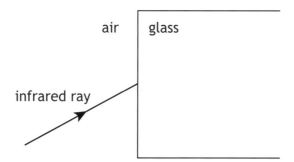

Complete the diagram to show the path of the infra-red ray as it enters the glass from air.

Indicate on your diagram the normal, the angle of incidence and the angle of refraction.

2

Total marks 8

MARKS

6. When welders join thick steel plates it is important that the joint is completely filled with metal. This ensures there are no air pockets in the metal weld, as this would weaken the joint.

One method of checking for air pockets is to use a radioactive source on one side of the joint. A detector placed as shown measures the count rate on the other side.

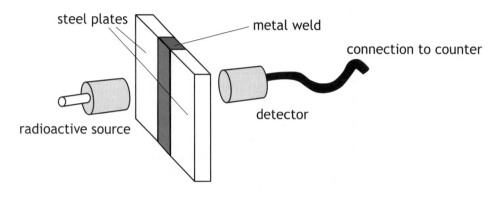

View from above

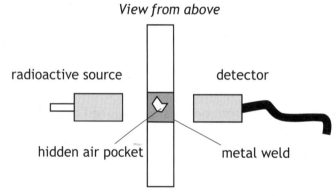

(a) The radioactive source and detector are moved along the weld.

How would the count rate change when the detector moves over an air pocket?

Explain your answer.   2

(b) Which of the radiations alpha, beta or gamma must be used?

Explain your answer.   2

Total marks 4

MARKS | DO NOT WRITE IN THIS MARGIN

7. (a) A medical physicist checks the count rate of a radioactive source. A graph of count rate against time for the source is shown. The count rate has been corrected for background radiation.

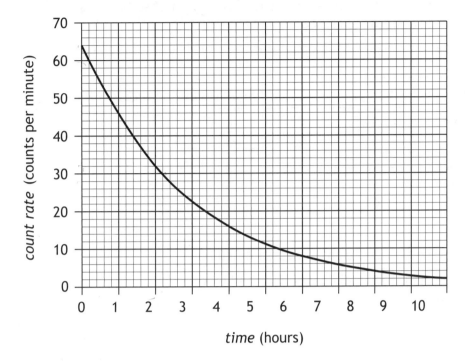

time (hours)

(i) Use the graph to determine the half-life of the source. 1

(ii) State **two** factors which can affect the background radiation level. 2

MARKS | DO NOT WRITE IN THIS MARGIN

7. (continued)

(b) Another medical physicist is investigating the effects of radiation on tissue samples. One sample of tissue receives an absorbed dose of 500 µGy of radiation from a source.

The radiation weighting factors of different types of radiation are shown.

Type of radiation	Radiation weighting factor (w_R)
gamma	1
thermal neutrons	3
fast neutrons	10
alpha	20

(i) The tissue sample has a mass of 0·040 kg.

Calculate the total energy absorbed by the tissue.   3

Space for working and answer.

(ii) The equivalent dose received for this tissue sample is 10·00 mSv.

Which type of radiation is the medical physicist using?

Justify your answer by calculation.   4

Total marks 10

MARKS | DO NOT WRITE IN THIS MARGIN

8.   A simple pendulum consists of a mass suspended on a string which is allowed to swing from a point. The pendulum oscillates back and forth when released.

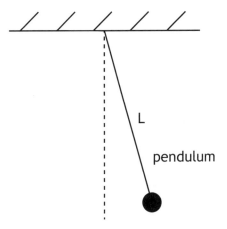

For small amplitudes, the period T of this pendulum can be approximated by the relationship:

$$T = 2\pi\sqrt{\dfrac{L}{g}}$$

Where: L is the length of the string (m).

g is the acceleration due to gravity (m s^{-2}).

(a)   Calculate the period of a pendulum with a string of length 0·8 m.   **2**

Space for working and answer.

MARKS | DO NOT WRITE IN THIS MARGIN

8. **(continued)**

(b) Calculate the frequency of the oscillation.   3
Space for working and answer.

(c) The mass of the pendulum in the above example is now doubled.

State the effect this has on the period of the pendulum.   1

Total marks   6

MARKS

9. The European Space Agency has been authorised to fly manned missions to the International Space Station (ISS).

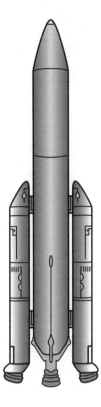

For one particular mission, a spacecraft with booster rockets attached will be launched.

(a) On the diagram above draw and label the two forces acting on the spacecraft at lift-off.

2

MARKS | DO NOT WRITE IN THIS MARGIN

9. (continued)

(b) The combined mass of the spacecraft and booster rockets is $3 \cdot 08 \times 10^5$ kg and the initial thrust on the rocket at lift-off is 3352 kN.

The frictional forces acting on the rocket at lift-off are negligible.

(i) Calculate the weight of the spacecraft and booster rockets at lift-off.   **3**

Space for working and answer.

(ii) Calculate the acceleration of the spacecraft and booster rockets at lift-off.   **4**

Space for working and answer.

MARKS

DO NOT WRITE IN THIS MARGIN

9. **(continued)**

(c) The ISS orbits at a height of approximately 360 km above the Earth.

Explain why the ISS stays in orbit around the Earth. 2

(d) An astronaut on board the ISS takes part in a video link-up with a group of students. The students see the astronaut floating.

(i) Explain why the astronaut appears to float. 1

(ii) The astronaut then pushes against a wall and moves off.

Explain in terms of Newton's Third Law why the astronaut moves. 2

Total marks 14

10. A car of mass 700 kg travels along a motorway at a constant speed. The driver sees a traffic hold-up ahead and performs an emergency stop. A graph of the car's motion is shown, from the moment the driver sees the hold-up.

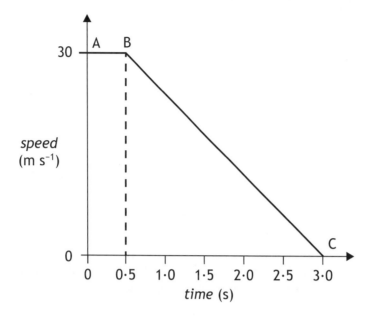

(a) Describe **and** explain the motion of the car between A and B.   2

MARKS

DO NOT WRITE IN THIS MARGIN

10. **(continued)**

(b) Calculate the kinetic energy of the car at A.   3

Space for working and answer.

(c) Show by calculation that the magnitude of the unbalanced force required to bring the car to a halt between B and C is 8400 N.   4

Space for working and answer.

Total marks   9

11. Some information about two racing cars is shown in the table.

Car	Maximum speed (m s^{-1})	maximum acceleration (m s^{-2})
A	40	4
B	20	9

The cars race on the following track.

Use your knowledge of physics to comment on which car would be most suitable for racing on this track.

3

MARKS | DO NOT WRITE IN THIS MARGIN

12. Read the passage below and answer the questions that follow.

Sunspots

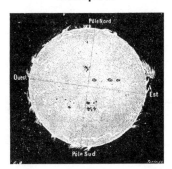

Sunspots are the dark spots which appear on the surface of the sun. They are caused by intense magnetic fields appearing beneath the sun's surface. These magnetic fields disrupt the natural convection within the sun, causing a reduction in the surface temperature at that point which appears as a dark spot.

Sunspots release of huge amount of energy known as 'solar flares'. These flares lead to the ejection of billions of tonnes of high energy solar particles. Electromagnetic (EM) radiation is also emitted. Such releases are known as coronal mass ejections (CME). The particles and energy travel into space.

High energy particles emitted by solar flares rarely reach Earth, and when they do, its magnetic field deflects most of them. Background radiation levels are not significantly changed by the particles which penetrate the atmosphere.

However, satellites in the path of these particles can suffer damage to sensitive equipment on board. Early warning of the arrival of solar flares allows operators to place satellites into 'protective mode'.

The arrival of high electromagnetic energy, in particular ultraviolet radiation and X-rays can also cause significant damage.

When this EM radiation reaches the Earth, it can cause changes to the upper atmosphere, which can affect the orbits of low altitude satellites. The EM radiation can also cause localized disturbances of the Earth's magnetic field. On one occasion, such interference was claimed to be responsible for causing power failures in Canada which affected millions of households for several hours.

The NASA space agency monitors sun spots, using their two satellite based Solar Terrestrial Relations Observatories (STEREO).

One particular observation from these satellites recorded a solar flare ejected from the sun which took 2·3 days to reach Mars, a distance of $2·28 \times 10^{11}$ metres from the sun.

Sunspot activity increases to a maximum in 11 year cycles.

There is on-going research to identify a connection between solar activity and our terrestrial climate.

Currently, scientists cannot predict when a sunspot or solar flare will appear, but detection and early warning techniques have improved.

MARKS

12. **(continued)**

(a) What causes sunspots to appear?

1

(b) Which parts of the electromagnetic spectrum can cause damage to satellites?

2

(c) Calculate the average speed in ms^{-1} of the solar flare which reached Mars.

Space for working and answer.

3

Total marks   6

[END OF MODEL QUESTION PAPER]

MARKS | DO NOT WRITE IN THIS MARGIN

ADDITIONAL SPACE FOR ROUGH WORKING AND ANSWERS

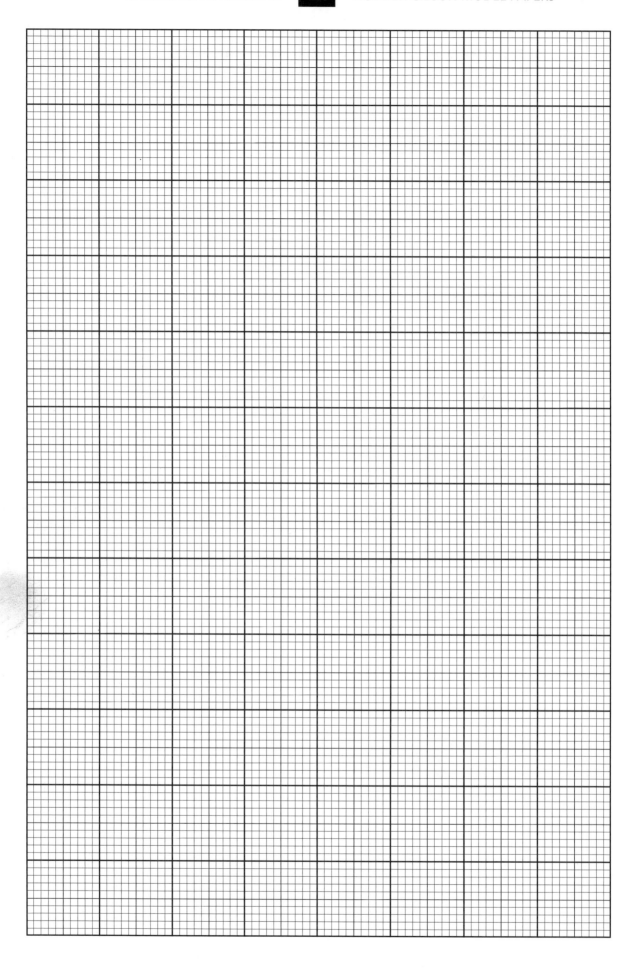

Section 1

1.	C	6.	C	11.	A	16.	D
2.	B	7.	E	12.	D	17.	A
3.	C	8.	B	13.	E	18.	D
4.	D	9.	C	14.	E	19.	D
5.	C	10.	C	15.	B	20.	A

Section 2

1. (a) (i) $R_1 = R_1 + R_2 + R_3$
   $$= (30 + 30 + 15 =) \ 75 (\Omega)$$
   $$I = \frac{V}{R}$$
   $$= \frac{15}{75}$$
   $$= 0.2 \ A$$

   (ii) $V = IR$
   $$= 15 \times 0.2$$
   $$= 3 \ V$$

   (b) Total circuit resistance is less so the reading on the ammeter will increase.

   Resistors in parallel:
   $$\frac{1}{R_t} = \frac{1}{R_1} + \frac{1}{R_2}$$
   $$\frac{1}{R_t} = \frac{1}{30} + \frac{1}{30}$$
   $$R_t = 15 \ \Omega$$

   Total resistance =
   $(15 + 15 =) \ 30 \ \Omega$

2. (a) $E_h = cm\Delta T$
   $$= 4180 \times 3 \times 30$$
   $$= 376 \ 200 \text{J}$$

   (b) $E_h = Pt$
   $376 \ 200 = 120 \times t$
   $t = 3135 \ s$

   (c) (i) Energy loss to surroundings

   (ii) Top open — use a cover/lid etc

3. This is an open-ended question.
   The maximum available mark would be awarded to a student who has demonstrated a good understanding of the physics involved. The student shows a good comprehension of the physics of the situation and has provided a logically correct answer to the question posed. This type of response might include a statement of the principles involved, a relationship or an equation, and the application of these to respond to the problem. This does not mean the answer has to be what might be termed an "excellent" answer or a "complete" one.

4. (a) Lower

   (b) House P: (rate of heat transfer)
   $$= 1.9 \times 300 \times 16$$
   $$= 9120 \ W$$

   House Q: (rate of heat transfer)
   $$= 0.6 \times 500 \times 16$$
   $$= 4800 \ W$$

   (c) Type B

   Type B glass transmits less infrared radiation than Type A glass.

5. (a)

Time before the stars join	Period of gravitational waves (s)	Frequency of gravitational waves (Hz)
1 million years	1000	0·001
1 second	0·0075 (1)	133 (1)
0·1 second	0·003 (1) or 0·0030	300 or (1) 330 or 333

   (b) The gravitational wave frequency increases.

   (c) $R = \frac{3.4 \times 10^8}{2} = 1.7 \times 10^8$

   $v = \frac{2 \times \pi}{T} R$

   $= \frac{2 \times \pi}{1150} \times 1.7 \times 10^8$

   $= 9.3 \times 10^5 \text{ m s}^{-1}$

6. (a)

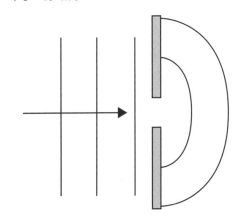

   (b) (i) Microwaves

   (ii) They all travel at the same speed through a vacuum OR in air.
   OR
   They all exhibit interference OR reflection OR refraction OR propagation.

7. (a) (i) Identify 13 Bq as half of the initial activity.
   Half-life is 5800 years.

   (ii) $26 \rightarrow 13 \rightarrow 6.5 = 2$ half-lives
   total time $= 2 \times 5800$
   $= 11\,600$ years

(iii) Activity of 125 mg sample is 40 Bq.
Activity of 8 mg of sample = 1/5
OR

$$\frac{25}{125} \times 40$$

$$= 8 \ (Bq)$$

(From graph, 8 Bq is at 9800 years.)

Sample is approximately 9800 years old.

(b) The half-life of carbon 14 is 5800 years.

For 100 years the very small reduction in the activity would be difficult to measure accurately.

8. (a) Gamma rays are electromagnetic waves.

(b) $f = \dfrac{v}{\lambda}$

$$= \frac{3 \times 10^8}{6 \times 10^{-13}}$$

$$= 5 \times 10^{20} \ \text{Hz}$$

(c) $\quad D = \dfrac{E}{m}$

$$50 \times 10^{-6} = \frac{E}{0 \cdot 1}$$

$$E = 5 \times 10^{-6} \ \text{J}$$

(d) Gamma rays are absorbed by the lead.

9. (a) $a = \dfrac{v - u}{t}$

$$= \frac{55 - 5}{40}$$

$$= 1 \cdot 25 \ \text{ms}^{-2} \ \text{Hz}$$

(b) (i)

North

Scale: 1 cm equivalent to 10 m s^{-1} (for example)

0

(ii) Change speed to 155 m s^{-1}
At bearing of 155 (or 15° West of North)

(c) aircraft has increased mass
so has reduced deceleration
OR
aircraft takes longer to stop
so longer distance required

10. (a) Total mass = 1300 + 2950 + 2900
$$= 7150 \ \text{kg}$$
$$F = ma$$
$$1430 = 7150 \times a$$
$$a = 0 \cdot 2 \ \text{ms}^{-2}$$

(b) (force of) friction (is created) on the surface of the modules

causes heat energy to be produced

(c) (i) upward force is increased (by parachutes)

producing an unbalanced force upward

(ii) $E_w = Fd$
$$80 \ 000 = F \times 5$$
$$F = 16 \ 000 \ \text{N}$$

11. (a) Milky Way

(b) $d = vt$
$$= 3 \times 10^8 \times (365 \times 24 \times 60 \times 60) \times 30000$$
$$= 2 \cdot 84 \times 19^{20} \ \text{m}$$

(c) Longer
Lower

(d) Helium
Hydrogen

12. This is an open ended question.
The maximum available mark would be awarded to a student who has demonstrated a good understanding of the physics involved. The student shows a good comprehension of the physics of the situation and has provided a logically correct answer to the question posed. This type of response might include a statement of the principles involved, a relationship or an equation, and the application of these to respond to the problem. This does not mean the answer has to be what might be termed an "excellent" answer or a "complete" one.

NATIONAL 5 PHYSICS MODEL PAPER 1

Section 1

1.	C	6.	C	11.	B	16.	B
2.	C	7.	A	12.	E	17.	D
3.	D	8.	B	13.	A	18.	D
4.	B	9.	E	14.	B	19.	A
5.	D	10.	A	15.	A	20.	C

Section 2

1. (a) $E_p = mgh$
   $= 235 \times 9 \cdot 8 \times 12$
   $= 27636$ J

   (b) (i) $E_p = E_K$

   $\frac{1}{2} mv^2 = mgh$

   $\frac{1}{2} \times 2 \cdot 5 \times v^2 = 2 \cdot 5 \times 9 \cdot 8 \times 12$

   $v = 15 \cdot 33$ m s^{-1}

   (ii) air resistance

2. (a) (i) $I = 0 \cdot 075$ A
      $V = IR$
      $4 \cdot 2 = 0 \cdot 075 \cdot R$
      $R = 56 \, \Omega$

   (ii) resistance stays the same because the graph is a straight line through the origin

   (b) $\frac{1}{R_1} = \frac{1}{R_1} + \frac{1}{R_2}$

   $= \frac{1}{33} + \frac{1}{56}$

   $= 0 \cdot 048$

   $R_t = 20 \cdot 76 \, \Omega$

3. (a) Terminal velocity occurs when forces acting on a moving object become balanced.

   (b) $F_d = 6\pi r\eta v_1$
   $= 6 \times \pi \times 2 \cdot 83 \times 10^{-6} \times 1 \cdot 820 \times 10^{-5} \times 8 \cdot 56 \times 10^{-4}$
   $= 8 \cdot 31 \times 10^{-13}$ N

   (c) $W = mg$
   $8 \cdot 6 \times 10^{-13} = m \times 9 \cdot 8$
   $m = 8 \cdot 8 \times 10^{-14}$ kg

   (d) a charged particle experiences a force in an electric field

4. (a) $E = cm\Delta T$
   $= 902 \times 8000 \times (660 - 160)$
   $= 3 \cdot 61 \times 10^9$ J

   (b) $l_f = 3 \cdot 95 \times 10^5$ J kg^{-1}
   $E = ml$
   $= 8000 \times 3 \cdot 95 \times 10^5$
   $= 3 \cdot 16 \times 10^9$ J

5. Assume that the average mass of a student is 60 kg.

   Student has a weight of $W = mg = 588$ N.

   The total weight of the student will be exerted on the ground where his feet are in contact.

   Assume that the area of one foot in contact with ground is $0 \cdot 2$ m $\times 0 \cdot 08$ m $= 0 \cdot 016$ m^2.

   Total area in contact $= 0 \cdot 032$ m^2.

   Pressure $= \frac{force}{area} = \frac{588}{0 \cdot 032} = 18375$ Pa

6. (a) $P \times V = 2000, 1995, 2002, 2001$

   $P \times V = $ constant

   (b) gas molecules collide with walls of container more often

   so average force increases

   causing an increase in pressure

   (c) To reduce the inaccuracy of the syringe volume, since the volume of air contained in the tubing is not measured.

7. (a) Refraction occurs when light travels from one medium into another with a change in the wave speed

   (b) (i)

   (ii)

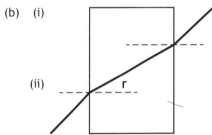

   (c) (i) P   ultraviolet
         Q   microwaves

   (ii) $d = vt$
   $4 \cdot 5 \times 10^{12} = 3 \times 10^8 \times t$
   $t = 1 \cdot 5 \times 10^4$ s

8. (a) A particle containing two protons and two neutrons

   (b) ionisation is the gain/loss of electrons by an atom

   (c) 4800 $\xrightarrow{1}$ 2400 $\xrightarrow{2}$ 1200 $\xrightarrow{3}$ 600 $\xrightarrow{4}$ 300

   4 half lives

   $4 \cdot 2 \cdot 5 = 10$ hours

   (d) Source may also emit β and/or γ radiation

9. (a) (i) $D = \frac{E}{m}$

   $= \frac{6 \times 10^{-6}}{0 \cdot 5}$

   $= 1 \cdot 2 \times 10^{-5}$ Gy

   (ii) $H = Dw_R$
   $= 1 \cdot 2 \times 10^{-5} \times 20$
   $= 2 \cdot 4 \times 10^{-4}$ Sv

   (iii) $A = \frac{N}{t}$

   $= \frac{24\,000}{(5 \times 60)}$

   $= 80$ Bq

   (b) Fission.

10. (a) (i)

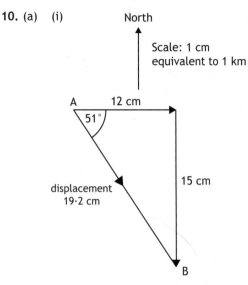

Scale: 1 cm equivalent to 1 km

A   12 cm

51°

15 cm

displacement
19·2 cm

B

Displacement is 19·2 km, bearing 141

(ii) $s = \bar{v}t$
    $19\cdot2 = \bar{v} \times 1\cdot25$
     $\bar{v} = 15\cdot4$ km h^{-1}, bearing 141

(b) Displacement is 19·2 km, bearing 141

11. (a) $a = \dfrac{v - u}{t}$

    $= \dfrac{3 - 0}{5}$

    $= 0\cdot6$ m s^{-2}

(b) $F = ma$
    $= 40 \times 0\cdot6$
    $= 24$ N

12. (a) (i) 1·1 MW

     (ii) power output not consistent

(b) different water speeds, different sizes of rotor blades

13. (a) Sagittarius A

(b) $d = vt$
    $2\cdot6 \times 10^{20} = 3 \times 10^{8} \times t$
    $t = 8\cdot7 \times 10^{11}$ seconds

    $= \dfrac{8\cdot7 \times 10^{11}}{365\cdot25 \times 24 \times 60 \times 60}$

    $= 27569$ *light years*

(c) (i) Geiger-muller tube

     (ii) Compared to gamma rays, light rays have a **lower** frequency which means they have a **lower** energy.

14. If air resistance is ignored, then at the planet surface all objects fall with the **same** acceleration due to gravity g, so objects will take the same time to fall equal distances.

Different planets have different values for this acceleration g. The time taken to fall the same distance would be longer on planets where g is smaller.

On Earth, air resistance could reduce the acceleration of a falling object. If it was not streamlined, then an object would take longer to fall to the ground.

NATIONAL 5 PHYSICS MODEL PAPER 2

Section 1

1.	E	6.	E	11.	D	16.	D
2.	B	7.	C	12.	E	17.	E
3.	C	8.	B	13.	A	18.	A
4.	B	9.	C	14.	E	19.	C
5.	C	10.	E	15.	D	20.	B

Section 2

1. (a) $E_p = mgh$

   $= 750 \times 9\cdot8 \times 7\cdot2$

   $= 52920$ J

   (b) (i) 52920 J

   (ii) $E_k = \frac{1}{2}mv^2$

   $52920 = \frac{1}{2} \times 750 \times v^2$

   $v = 11\cdot9$ m s^{-1}

2. (a) the resistance of LDR decreases when light level rises

   voltage across R rises until MOSFET switches on the motor

   (b) (i) 3000 Ω

   (ii) $V_1 = \left(\frac{R_1}{R_1 + R_2}\right) V_s$

   $V_1 = \left(\frac{600}{600 + 3000}\right) \times 12$

   $V_1 = 2$ V

   (iii) Since V is less than 2·4 V the transistor will not switch on so blinds do not shut.

3. (a) pressure is the force per unit area exerted on a surface

   (b) $p = \rho gh$

   $= 1025 \times 9\cdot8 \times 24$

   $= 2\cdot4 \times 10^5$ Pa

4. Cars have to speed up regularly, reducing the overall mass of the car would reduce the unbalanced force required for acceleration (F=ma) reducing fuel needed.

   Making the car more streamlined would reduce the frictional forces acting on the car, and so less fuel would be needed to provide the reduced forward force.

   Design cars which transform their kinetic energy into electrical energy when braking, to be stored in rechargeable cells. These cells can provide an additional energy source to energise a motor which could assist the car's movement.

5. (a) $\frac{p}{T}$ = 347, 347, 346, 348, 348

   pressure and temperature are directly proportional when T is in Kelvin

   (b) As temperature increases, the average E_k of the gas particles increases, so the particles collide with the walls of the container more frequently with greater force so pressure increases.

   (c) To ensure all the gas in the flask is heated evenly

6. (a) (i) $v = f\lambda$

   $f = \frac{3 \times 10^8}{0\cdot06}$

   $= 5 \times 10^9$ Hz

   (ii) $T = \frac{1}{f}$

   $= \frac{1}{5 \times 10^9}$

   $= 2 \times 10^{-10}$ s

   (b) signals received at same time
   radio waves and microwaves have same speed

   (c) The radio signals for car B have much higher frequency than for car A, so the radio signals for car B do not diffract as much around hills

7. (a) half-life is the time for the activity of a radioactive substance to reduce to half of its original value

   (b) *no. of half lives* $= \frac{\textit{no. of days}}{\textit{half life}} = \frac{13\cdot5}{2\cdot7} = 5$

   64 ⟨1⟩ 32 ⟨2⟩ 16 ⟨3⟩ 8 ⟨4⟩ 4 ⟨5⟩ 2

   activity after 13·5 days = 2 kBq

   (c) minimise the time handling the sources, keep the radiation source at a maximum distance from the body.

8. (a) in a fission reaction, a nucleus with a large mass number splits into two nuclei of smaller mass numbers

   energy is released and neutrons are usually released

   (b) $P = \frac{E}{t}$

   $E = 1\cdot4 \times 10^9 \times 60 \times 60$

   $E = 5\cdot0 \times 10^{12}$ (J)

   Number of fissions $= \frac{5\cdot0 \times 10^{12}}{2\cdot9 \times 10^{-11}}$

   $= 1\cdot7 \times 10^{23}$

   (c) radioactive waste requires expensive storage

9. (a) $W = mg$

   $= 50\,000 \times 9\cdot8$

   $= 4\cdot9 \times 10^5$ N

   (b) $4\cdot9 \times 10^5$ N

   (c)

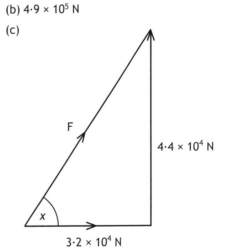

   $F^2 = (3\cdot2 \times 10^4)^2 + (4\cdot4 \times 10^4)^2$

   $F = 5\cdot4 \times 10^4$ N

   $\tan x = \frac{4\cdot4 \times 10^4}{3\cdot2 \times 10^4}$

   $x = 54°$

   resultant force = $5\cdot4 \times 10^4$ N at bearing of 036

(d) $H = Dw_R$

$\quad = 15 \times 10^{-6} \times 1$

$\quad = 1 \cdot 5 \times 10^{-5} \ Sv$

(e) ionisation is when an atom gains or loses electrons

10. (a) $a = \dfrac{v - u}{t}$

$\quad = \dfrac{9 - 0}{2}$

$a = 4 \cdot 5 \ m \ s^{-2}$

(b) $F = ma$

$\quad = 15 \times 4 \cdot 5$

$\quad = 67 \cdot 5 \ N$

(c) d = area under graph

$\quad = (0 \cdot 5 \times 9 \times 2) + (10 \times 9) + (0 \cdot 5 \times 9 \times 1)$

$\quad = 9 + 90 + 4 \cdot 5$

$\quad = 103 \cdot 5m$

11. (a) $E_w = F \times d$

$\quad = 300 \times 1 \cdot 5 \times 500$

$\quad = 225\,000 \ J$

(b) (i) $E = c \ m \ \Delta T$

$\quad 225\,000 = 902 \times 12 \times \Delta T$

$\quad \Delta T = 21 \ ^\circ C$

(ii) energy is lost to the surrounding air

12. As the rocket rises, fuel is used up and the rocket mass reduces. This reduces the rocket weight and so the unbalanced upward force increases, causing acceleration to increase as $a = \dfrac{F}{m}$.

As the rocket rises, the gravitational field strength decreases. This causes the rocket weight to reduce and so the unbalanced upward force increases, causing acceleration to increase as $a = \dfrac{F}{m}$.

As the rocket rises, the frictional force reduces because of the atmosphere being less dense at altitude, this reduces air resistance and so the unbalanced upward force increases, causing acceleration to increase as $a = \dfrac{F}{m}$.

13. (a) Neutron stars are thought to be formed when large stars collapse.

(b) During the collapsing process, electrons and protons combine to form neutrons.

(c) $T = \dfrac{1}{f}$

$\quad = \dfrac{1}{716}$

$\quad = 1 \cdot 4 \times 10^{-3} \ Hz$

NATIONAL 5 PHYSICS MODEL PAPER 3

Section 1

1.	B	6.	A	11.	C	16.	D
2.	D	7.	C	12.	E	17.	E
3.	D	8.	C	13.	A	18.	D
4.	C	9.	E	14.	D	19.	C
5.	E	10.	A	15.	B	20.	E

Section 2

1. (a) $E_p = mgh$

$\quad = 0 \cdot 50 \times 9 \cdot 8 \times 19.3$

$\quad = 94.6 \ J$

(b) $Eh = cm\Delta T$

$\quad 94 \cdot 6 = 386 \times 0 \cdot 50 \times \Delta T$

$\quad \Delta T = 0 \cdot 49 \ ^\circ C$

(c) Temperature change is less.

Some heat energy is lost to the surroundings.

2. (a) $P = I^2 R$

$\quad 2 = I^2 \times 50$

$\quad I^2 = 0 \cdot 04$

$\quad I = 0 \cdot 2 \ A$

(b) (i) $\dfrac{1}{R_T} = \dfrac{1}{R_1} + \dfrac{1}{R_2}$

$\quad \dfrac{1}{R_T} = \dfrac{1}{60} + \dfrac{1}{30}$

$\quad R_T = 20 \ \Omega$

(ii) $P = \dfrac{V^2}{R}$

$\quad P = \dfrac{9^2}{60}$

$\quad = 1 \cdot 35 \ W$

$\quad P = \dfrac{V^2}{R}$

$\quad P = \dfrac{9^2}{30}$

$\quad = 2 \cdot 7 \ W$

(iii) 30 Ω resistor will overheat

(c) none

3. The final temperature will be between both starting temperatures.

The heat energy E_h lost by the copper will be gained by the water.

From $\Delta T = \dfrac{E_h}{cm}$ the rise in temperature of the water and fall in temperature copper will depend on the respective masses of water and copper.

Specific heat capacity for water is much greater than for copper, so if masses were equal, final temperature would be closer to water's starting temperature.

4. (a) $p = \dfrac{F}{A}$

   $F = 4\cdot6 \times 10^5 \times 3\cdot00 \times 10^{-2}$

   $F = 13800$ N

   (b) $P_1V_1 = P_2V_2$

   $4\cdot6 \times 10^5 \times 1\cdot6 \times 10^{-3} = 1\cdot0 \times 10^5 \times V_2$

   $V_2 = 7\cdot36 \times 10^{-3}$ m^3

   (c) $V_{water} = V_2 - V_1 = 7\cdot36 \times 10^{-3} - 1\cdot6 \times 10^{-3} = 5\cdot76 \times 10^{-3}$ m^3

5. (a) $\lambda = \dfrac{distance}{no.\ of\ waves} = \dfrac{36}{8} = 4\cdot5$ cm $= 0\cdot045$ m

   $v = f\lambda$

   $= 5 \times 0\cdot045$

   $= 0\cdot23$ m s^{-1}

   (b)
   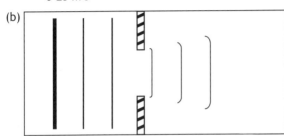

   (c)
   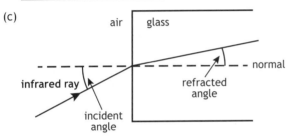

6. (a) count rate increases

   less metal to be penetrated

   (b) gamma

   penetrates the metal best

7. (a) (i) half-life = 2 hours

   (ii) cosmic rays, fall out from nuclear accidents

   (b) (i) $D = \dfrac{E}{m}$

   $500 \times 10^{-6} = \dfrac{E}{0\cdot04}$

   $E = 2 \times 10^{-5}$ J

   (ii) $H = Dw_R$

   $10 \times 10^{-3} = 500 \times 10^{-6} \times w_R$

   $w_R = 20$

   alpha radiation

8. (a) $T = 2\pi \sqrt{\dfrac{L}{g}}$

   $= 2\pi \sqrt{\dfrac{0\cdot8}{9\cdot8}}$

   $= 1\cdot8$ s

   (b) $T = \dfrac{1}{f}$

   $1\cdot8 = \dfrac{1}{f}$

   $f = 0\cdot6$ Hz

   (c) no effect

9. (a)
   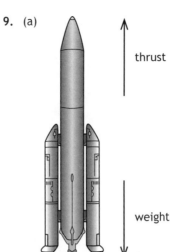

   (b) (i) $W = mg$

   $= 3\cdot08 \times 10^5 \times 9\cdot8$

   $= 3\cdot02 \times 10^6$ N

   (ii) $F_{unbalanced} = 3352 \times 10^3 - 3\cdot02 \times 10^6 = 332\,000$ N

   $F = ma$

   $332\,000 = 3\cdot08 \times 10^5 \times a$

   $a = 1\cdot08$ m s^{-2}

   (c) the ISS moves with constant speed in the horizontal direction

   while accelerating due to the force of gravity in the vertical direction

   (d) (i) The astronaut is falling towards Earth at the same rate as the ISS

   (ii) the astronaut exerts a force against the wall

   the wall exerts an equal and opposite force against the astronaut causing him to move

10. (a) car continues at a <u>constant speed</u> during this time

    AB represents driver's reaction time

    (b) $E_k = \dfrac{1}{2}mv^2$

    $= 0\cdot5 \times 700 \times 30^2$

    $= 315\,000$ J

    (c) $a = \dfrac{v - u}{t}$

    $= \dfrac{0 - 30}{2\cdot5}$

    $= -12$ m s^{-2} = deceleration

    $F = ma$

    $= 700 \times 12$

    $= 8400$ N

11. racetrack has many bends and changes of direction which require both cars to decelerate and accelerate often

    car B has highest acceleration which would allow it to reach higher speeds more quickly after a bend

    the higher maximum speed of car A is unlikely to be used because of the limited number of long straight parts of track before decelerating is required.

12. (a) Sunspots are caused by intense magnetic fields appearing beneath the sun's surface.

(b) ultraviolet radiation and X-rays

(c) $d = \bar{v}\, t$

$2 \cdot 28 \times 10^{11} = \bar{v} \times 2 \cdot 3 \times 24 \times 60 \times 60$

$\bar{v} = 1 \cdot 15 \times 10^{6}\ \text{m s}^{-1}$

Acknowledgements

Permission has been sought from all relevant copyright holders and Hodder Gibson is grateful for the use of the following:

Image © NASA/JPL-Caltech (SQP Section 1 page 12);

Image © GSFC/D.Berry (SQP Section 2 page 13);

Image © NRC-CNRC (Harry Turner) (SQP Section 2 page 18);

An extract from "Dragonfish nebula conceals giant star cluster" taken from the 'New Scientist Magazine', 26 January 2011 © 2011 Reed Business Information - UK. All rights reserved. Distributed by Tribune Media Services (SQP Section 2 page 26);

Image © NASA/JPL-Caltech/Univ. of Toronto (SQP Section 2 page 26);

Image © Edwin Verin/Shutterstock.com (Model Paper 1 Section 1 page 10);

Image © jupeart/Shutterstock.com (Model Paper 1 Section 2 page 25);

Image © Joggie Botma/Shutterstock.com (Model Paper 1 Section 2 page 27);

Image © Rich Carey/Shutterstock.com (Model Paper 2 Section 2 page 8);

Image © Alexander Gordeyev/Shutterstock.com (Model Paper 2 Section 2 page 23);

Image © LingHK/Shutterstock.com (Model Paper 2 Section 2 page 24);

Image © Morphart Creation/Shutterstock.com (Model Paper 3 Section 2 page 25).

Hodder Gibson would like to thank SQA for use of any past exam questions that may have been used in model papers, whether amended or in original form.